SCARCITY'S WAYS:
THE ORIGINS OF CAPITAL

BOSTON STUDIES IN THE PHILOSOPHY OF SCIENCE

VOLUME 176

MICHAEL S. MACRAKIS

SCARCITY'S WAYS: THE ORIGINS OF CAPITAL

A Critical Essay on Thermodynamics, Statistical Mechanics and Economics

SPRINGER-SCIENCE+BUSINESS MEDIA, B.V.

A C.I.P. Catalogue record for this book is available from the Library of Congress

ISBN 978-90-481-4919-3 ISBN 978-94-015-8861-4 (eBook)
DOI 10.1007/978-94-015-8861-4

Printed on acid-free paper

Dedicated to Lily, Stavros,
Kristie and Michele

TABLE OF CONTENTS

I	PREFACE & INTRODUCTION	1
II	ECONOMICS	17
	1. Capital Theory	18
	2. Whence Capital	33
	3. La Raison Dialectique	40
	4. Summary	52
	5. A Note on Philosophical Terms	54
III	BIOLOGY	61
	1. Introduction, Biology 101	61
	[The Economics of Biology]	
	2. Biology 201	72
	[The Dialectics of Biology]	87
	3. Philosophical Issues	87
	4. Conclusions	88
IV	THERMODYNAMICS ($\Theta\Delta^{CS}$), STATISTICAL MECHANICS, AND CAPITAL	93
	1. Introduction	93
	2. Nephelokokkygia #1	94
	3. Nephelokokkygia #2	104
	4. Intermezzo	117

V THERMODYNAMICS ($\Theta\Delta^{CS}$), WORK, AND CAPITAL 125

1. Work, all Work and Nothing but Work 125
2. Biological Work, Microscopic Engines 138
3. Measurement in Physics 151

VI SUMMARY AND CONCLUSIONS 173

VII APPENDICES 183

1. On Thermodynamics and Statistic Mechanics 185
2. Availability 191
3. Thermoeconomics 194
4. Tractatus de Signis. Utilitarian Concerns and Scarcity 196
5. Prolegomena on Measurement 206

VIII BIBLIOGRAPHY 215

INDEX 223

List of Figures

II-1	The Structure of an Input-Output table	21
II-2	Philosophical Schools	57
II-3	Past and Future	59
III-1	Lwoff's energy flows in biological systems	60
III-2	An elementary Input-Output table biological I-O table	65
III-3	The *Central Dogma* illustrated	67
III-4	Towards the development of a biological I-O table	70
IV-1	A reductionist, meliorative program	106
IV-2	Evolution of Structures	119
V-1	(a) Heat transfer at an interface and (b) the Carnot Engine with a shaft for work	132
V-2	The Steam-Water and the ATP-ADP cycles	140
V-3	Huxley's Z-Plates	142
V-4	The Szilard box	146
V-5	The LEED set-up (schematic)	164
V-6	The Stern-Gerlach set-up	166
VI-1	U for UNIVERSE	172
A-1	The grand canonical ensemble. Region II	189
A-2	Experimental Categories	208-9

CHAPTER I

PREFACE & INTRODUCTION

THE PRINCIPAL ISSUES ADDRESSED in this exploratory essay are the *Origins of Capital* and the *Foundations of Thermodynamics*.

In the economic literature the existence of capital is taken for granted yet it is difficult to establish a robust definition of capital and to circumscribe its functions. Capital theory — the principal subject of Political Economy — is a theory which concerns itself with the coupling of the present with the future; as such it does not deal with its own history, its evolution or its origins. Yet, in its generic or reference form, capital is nothing but a physical or biological engine which processes materials and transforms energy in a physical environment of thermal non-equilibrium and of resource constraints. The objective of this essay is to connect the two themes of capital and thermodynamics and show the tight relation between the question of the origin of capital and the essentiality of capital in the comprehension of thermodynamics.

To attain the objectives and support my position, a review of capital theory — as presently understood — of evolutionary biology and the origins of life, and of the received ideas on thermodynamics became essential. The motivation and the mechanics for this ambitious program deserve some introductory remarks.[1]

The subject matter of this essay has been worked and reworked over the last several years as I have been puzzling first about problems concerning the fundamental issues of statistical mechanics, and then about problems relating to energy policy, economics and the thermodynamics of energy conservation. These inquiries originated and advanced in quite different settings and environments and involved various professional fields normally considered totally decoupled. Yet their junctions have been the inspiration and the justification of this essay.

Some twenty odd years ago I was working in various aspects of plasma physics and in particular the theory of interaction of electromagnetic waves with matter and its fluctuations. The field stands on the study of transport theory, statistical mechanics, and the thermodynamics of many-body systems. But when one scratches the integument of these beautiful edifices by studying the literature, deep questions arise as to the legitimacy of thermodynamics and of statistical mechanics in addressing the puzzle of irreversibility. It is indeed difficult to make up one's mind whether some of this work is a work of art or the result of calculation and derivation. With legerdemain the various authors seem to make incursions into the "microscopics" of the many-body problem through the two-, three-, and the n-th particle correlation functions and then switch to the macroscopics of temperature

[1] In what follows an author's name followed by a parenthesis with year of publication and often a page number refers to the bibliography at the end of the book.

invoking the fluctuation-dissipation theorems of Einstein and more recently those of Callen and Kubo[2] in order to get to manageable results. In this manner, great strides have been made in the development of the many-body problem without reaching the necessity to legitimize thermodynamics. Thermodynamics and statistical mechanics were used successfully as bridges and guides to get the new ideas to conform to the macroscopic experiences (measurements). Hence the disinterest of theoreticians.[3] The frustration in attracting the attention of the working physicists on this problem is vividly described in Carnap's (1978) account on the reception of his ideas and efforts at the Institute of Advanced Studies at Princeton where he worked on the *Two Essays on Entropy*; it almost amounted, he thought, to a conspiracy of silence. The priorities of theoretical physicists remain with the puzzles and the cranking at hand: Quantum Field Theory (QFT) and "creation and annihilation" physics, solid state physics, — to mention only a few research areas — command the attention of those working on the research frontiers. Thermodynamics is taken for granted and is thought to be an almost depleted research area. Whatever the subterfuges, the hand-waving arguments, the paedagogical red herrings, the procedures work and nothing can be gained by allocating intellectual resources to resolve ambiguitics with few, if any, expected rewards. Down deep, all believe that the regularization of the field will one day be accomplished. In the meantime there is

---

[2] The fluctuation-dissipation theorem is a powerful example of the technique connecting the energy of vibrational modes to the assignment of an energy kT per mode, where T is the absolute temperature and k is Boltzmann's constant (1.38054×10^{-23}J°K^{-1}). For a review see the article of R. Kubo, "The Fluctuation-Dissipation Theorem," *Reports on Progress in Physics,* XXIX, 1966: 255-284.

[3] See further comments in Chapters III and IV; also O. Penrose (1990).

really no doubt whatsoever that intuition depicts physical reality adequately.[4] In a satisficing way.

The questioning that led to this essay, however, came from a different experience: from an inquiry into matters of energy and economics rather than from research in physics. Early in the 70s, I had been involved with questions of energy and of energy policy and I felt the need to study economic theory in order to be able to communicate with those specialists working in the fields of economic growth, of resources, and their optimal depletion. One of my first inquiries was in connection with the functional relation of *capital*, K, (rather than energy, E, as was then the vogue) with GNP (Gross National Product). While, as we shall see in chapter II, capital and its measurement are notoriously ladden with ideology and misconceptions, I ventured a preliminary correlation of capital with energy by using U.S. data on capital from the celebrated paper of Solow (1957), and time series for energy use; the data showed a "linear" dependence of K and E during considerable stretches of time. A more sophisticated analysis indicated a strong and surprising *complementarity* between K and E in the (collective) manufacturing sectors,[5] a finding

[4] The picture changed somewhat with the work of Prigogine and his associates (George, et al. 1972, 1979) who promised a resolution of the riddle of irreversibility by relegating the problem to a "superoperator" (or, the latter day Demon) acting directly on the microscopic quantities; but one is left with lingering doubts.

[5] I refer to this work more extensively in the text. I note here, however, that the findings on complementarity have not remained unchallenged, as they go against the grain of every economic thinking about matters of substitution — the cornerstone of economics. With a variation on the demand for capital ΔK at a price p_K and of energy ΔE at p_E, the quantity $\varepsilon_{EK} = p_E \partial K / p_K \partial E$ can be positive or negative and, accordingly indicative of substitution or complementarity. Complementarity means that if the price of energy increases, not only energy use will go down ($\partial E<0$) as one might expect, but also the capital used. It has been a controversial subject which elicited many research papers but it was conceded by R.S. Pindyck and J.J. Rotenberg,

which challenges the engineering-thermodynamic expectations of microeconomic subtitutability between K and E and which was not universally accepted by econometricians.

In general, energy fits in the field of Resource Economics which deals with the intertemporal management of depletion, with prices, elasticity of demand, and substitution. Any attempt, however, to bring the *physical* aspects of energy into economics, except through the microeconomic association of cost of the resource (which may include externalities), the cost of capital and its maintenance in use, is bound to meet with resistance, if not outright hostility and derision, from the professional economist.

True, many of the tools of economics have been borrowed from the physical sciences,[6] yet the contacts of economics with the physical sciences remain tenuous. The economic process itself *is* clearly an entropic one and Georgescu-Rœgen (1971), an economist, has attempted to bring the disciplines together but the effort only resulted into an erudite book without issue. The difficulties appear not so much because of the weak connections between fields, but rather because each subdiscipline has its own priorities and agendas

---

"Dynamic Factor Demands and the effect of Energy Price Shocks," *The Am. Ec. Rev.*, **73**, 3, 1983. The findings of complementarity in K-E may also be taken to support the *structuralist schools* (see discussion in S. Marglin, "The Wealth of Nations," *The New York Review of Books*, 19 July 1984).

[6] The seminal work of P.A. Samuelson, *Foundations of Economic Analysis,* (1948), abounds with the mathematics of thermodynamics especially those introduced by Maxwell transferred to economics. On the other hand the mathematics of the Fermat principle and the Calculus of Variations have been carried over to growth economics with great success. Note that identity in formal mathematical expressions does not make for good parallels among fields. The subtitle of my essay combines two physical disciplines with economics so that the reader must be warned not to read any attempts at "unification" of fields.

defining what is thought to be important and its own unwritten sign codes and short hands. Practitioners of accepted disciplines are invariably uneasy with interdisciplinary[7] works and endeavors. Schumpeter expressed this as the equivalent of the Monroe doctrine:[8] no tresspassing allowed even to adjacent disciplines. Apparently this is an indispensable guideline especially useful in a field where, as in physics, isolation of an effect from interactions is the key to scientific progress. Interestingly, however, some of the concepts in physics do have economic-looking content (the seeking of minimum paths for instance) but these, in turn, have stimulated the most advanced (positive) techniques in the study of economic growth thus supplementing the normative. Economics may have inspired the first trigonometers but modern mathematical economics has borrowed its techniques from physics, engineering, and mathematics.

Economists also practice the art of rhetoric, the better ones superbly. The art consists of articulating what in their minds appears as mathematical equations with very complex mathematical interrelationships — juggling, as it were, many inter-

---

[7] Interdisciplinary efforts are in general unproductive and deceiving. The latter because they give the impression of authoritativeness and completeness when in fact they may be collages of dubious merit. Occasionally, however, they can bear fruit. As Einstein wrote to Solovine, Poincaré's *La Science et l'Hypothèse* "profoundly impressed us and kept us breathless for weeks on end," Pais (1990: 134) quoting A. Einstein, *Lettres à Maurice Solovine,* p. VIII, Gauthiers-Villars, Paris 1956. Recall too that the celebrated essay of Schrödinger *What is Life?* had a similar stimulant effect on scores of biologists by constraining the possible mechanisms of reproduction. A similar stimulant is now needed to tackle the functioning of the brain as a conscious mechanism rather than as a simple machine element in the interaction loop.

[8] See R. Swedberg, *Schumpeter, A Biography,* Princeton University Press, 1991: 27.

related items with precision.[9] But even the simplest economic notion of "marginal," a short hand for the mathematical derivative that originated in physics, takes heaps of language to describe, when, in fact, a short course in calculus would have been a far better investment. But this is a way of reaching and influencing those in politics who wish to develop policies. Probably a difficult approach even for graduate students who prefer an equation to its verbalization. Were physicists to depend on such methods they would never have made much progress. In their work "simple" basics such as Newtonian or relativistic equations, quantum equations, and the like, obtain their power from the algorithmics and predictive measurables they lead to. The art is that of simplification of equations in which the utmost confidence is placed and which ingenuity and broad training in diverse areas of mathematics helps to solve. The rhetoric exists here too but it is much reduced.[10] This is the key of working the "paradigm." In economics, on the other hand, the equations themselves are not representationally robust or complete and neither do their users suggest that they have deeper roots. It is, then, the need to resolve these interdisciplinary problems which prompted me to explore extensively and intensively various field in search of clues; this book is the result of this Odyssey. The material is presented in the form of an exploratory essay rather than in the form of a technical paper; the latter could only justify its existence by addressing a well

---

[9] Part of the art of the economist is good "rhetoric" by which I mean the art of articulation of complex mathematical economic relationships without as much as a hint of the underlying mathematical grid; to this must be added the art of deriving mathematical expressions depicting the economic process by using intuition and "economic logic." See also ftn. 37 of Chapter II.

[10] This approach does not relate to the inventive aspects of science in which new paths and new paradigms on the fundamentals are forged. In this domain the rhetoric is often couched under subtle jumps in logic.

defined problem, by using well defined and widely accepted or discipline-centered methodological tools, and with the objective of reaching a novel conclusion.

With an exploratory essay then, I allow myself the freedom to study, identify, address a problem, without commitment to generate a constrained answer and I take solace from Quine's judgement that it "is not that everything worth establishing can be translated into the technical vocabulary of physics."[11] It is an attempt towards a synthesis and a positioning of a problem; an assembly of fragments of theories and of experiments from diverse disciplines to address the question of the origin of capital in its broadest sense. In the process, an unwieldy set of references resulted. From the extensive bibliography consulted only those items which were judged to be seminal or paedagogically useful were retained. One should recognize that the literature on the subjects discussed is indeed enormous. Just think that in his review article "General Properties of Entropy" A. Wehrl (1978) lists over 500 entries, each with some professional importance. O. Penrose (1979), in his review of the foundations of statistical mechanics, lists another, and different, 500 entries. Consulting these, branched to still more references. Perhaps this suggests a measure of the difficulty to contribute original and seminal material at the margin and also the questionable usefulness of comprehensive but uncritical reviews without a historical background. It soon became clear that not all material could be studied in any degree of detail; some articles, in addition, required special and current preparation for their reading and comprehension. The task was indeed too large

[11] W.V. Quine, "Otherwordly," a book review in *The New York Review of Books*, 23 November 1978. See also ftn. 3 of Chapter VI.

and risky. "I can testify from personal experience," commented Koopmans, "about the obstacles encountered by one trained in another field [physics] who embarks on the study of economics and seeks to absorb its substance."[12]

Almost autobiographically, I discussed the two fundamental research areas which remained disjointed in my mind for a considerable time. But the final justification for the essay and the proposed thesis came when I asked the question: *Whence Capital*? This question has opened up the path for a legitimate inquiry. One aspect of the origins of capital relates to the issues of the impact of the dynamics of capital accumulation on industrialization. But this only interests us here tangentially; it is a question that has preoccupied Marx and which Gerschenkron discusses extensively[13] with a strong focus on economic development and not without bite. Discounting various sources of original capital accumulation — Drake's booty, piracy, and precious metals — he refers to (p. 33) "an accumulation of capital continuing over long historical periods — perhaps over several centuries — until one day the tocsin of the industrial revolution was to summon it to the battlefields of factory construction." Then, elaborating on the "spurt in industrialization" (p. 99), he continues by stating that "it is this assumption which makes the concept of the 'beginning' a meaningful one, and it is the large amounts of capital needed to launch and sustain such a spurt that alone

---

[12] Koopmans (1957: 145) began his carrier as a physicist with important contributions in quantum mechanical subjects and shifted to economics, the field in which he *also* did seminal work that earned him the Nobel prize a 1975.

[13] Alexander Gerschenkron, *Economic Backwardness in Historical Perspective,* Harvard University Press, 1962. One is tempted to quote long passages from this book but the interested reader should refer to the source. See also p. 37 in the following chapter.

justify the concept of original, that is, prespurt accumulation. Without the industrial spurt the concept is destitute of meaning." He believes that the question as to where the 'first capital' came from is hardly sensible or interesting. But, in practice, economists do conceive the economy as immortal and, in their *equations of motion*, invariably allow for the pre-existence of capital at $t = -\infty$. They circumvent this defect by beginning their story saying: "by date t_{in}, the economy will have inherited a certain stock of capital and other conditions from its past behaviour." In this manner they make a cut in the historical process looking only into the future except, perhaps, when they retrodict. Indeed their concern is the (discounted) future rather than the historical or even prehistorical past; for them, at best, the Hesiodian economy marks the beginning of economic history; Hesiodic man is of the Iron Age. During the Golden Age which preceded it, men "lived like gods," without old age or real death. "All goods were theirs: the life-giving earth (ζείδωρος ἄρουρα) yielded, on its own, an abundant and generous produce," according to Hesiod (op. 173). Scarcity did appear with the Age of Iron! Again economic capital did not always exist. It was preceded by biological and physical "capital" — the structural assembly which allows for coherent transactions of energy. It must have started out from "nowhere." This is the beginning of my inquiry.

I felt too, that the subject deserved a solid philosophical backing. Yet the philosophical aspects have been the most difficult and vexing to express; for not only philosophy is not my field (as is also true with economics and biology) but the rigors of acquiring a perspective in a field as long-standing as philosophy proved unbearably large. Nevertheless, I may have inadvertently encountered (stumbled into might be more appropriate) the important thinking bearing on the objectives of this essay. As to the sequencing of the material I note that while a leading question has been the legitimacy of thermodynamics in terms of "first principles," the thrust of the propositions in the essay originated from economics; this suggested that I should begin chapter II with Economics and review Capital and Production Theory in a way to fit the needs of the essay. There, I stress the issues of scarcity — the theme that suggested the title of the book — and the political, economic, and philosophical questions pertaining to the origin of Capital. In chapter III, I discuss the links connecting biological structures with capital theory.[14] Finally, in chapters IV and V, I discuss the implications of the earlier chapters on the development of thermodynamics out of first principles; as the section titles show, a number of items are brought forward for discussion in order to support my case. Many of these relate to still unsolved problems such as decoherence or consciousness but they surely are relevant to keep in mind. Inevitably, Conclusions (VI) follow where I summarize my positions. To avoid loading the manuscript with technicalities I have relegated a few items to appendices. One is a rapid overview of Thermodynamics and Statistical Mechanics in three sections, in an attempt to look into accepted practice through the point of view and needs of the present essay. Finally, in the appendices, and *in lieu* of long footnotes, I develop two short pieces on "information" (being a critique on the "information theory of value") and "prolegomena" to measurement; they should be considered

---

[14] In fact the temptation was to entitle this book "Disease of Matter" which is what an uncle of Beckett called life: Knowlson, James, *The Life of Samuel Beckett*, Simon & Schuster 1996.

as descriptions of research agendas rather than as results of research accomplished.[15]

In the process of this research some new ideas have emerged. One of them is the biological input-output table where I construct a formal table that can be used as a guide to the mechanics of capital formation in biological systems stressing the roundaboutness of the process. I, then, introduce some new terms such as the *Keynesion,* the *Smithion, quasicapital, virtual capital, virtual Cartesian observer, interutility, biological input-output table, diergodicity, information theory of value* —the result of transferring concepts and techniques from one field to another.

Throughout I could not evade being confronted with some fundamental problems that bedevil physics. Not a few books have appeared recently that might reasonably be entitled *Critique of the Quantum Enigma.* No matter where one turns, one finds that the edifices on which modern science is founded are of questionable robustness. It is astonishing, nevertheless, that so much has been gleaned out of such edifice. We are not anywhere closer than Aristotle to having answers as to the "what it is that we are" and "where we fit" in the cosmos. But, amazingly, we do have better and precise metaphors and procedures to manipulate nature as it is. Science has not remained immune to the "postmodernist" introspection and uncertainty that has shaken the humanities over the last half century. While the historians of science had earlier written their findings in a narrative form and philoso-

---

[15] The material on information has been taken from the forthcoming *Causes for the Destabilization of the U.S. Telecommunications Monopoly — A Historical and Critical Perspective*; that on measurement from a proposal to NSF of 1995.

phers of science have been following the traditional analysis, one now observes a critical stance that gives the appearance of wanting to bring the edifices down. Reacting, the scientists give the impression of being threatened when, first in disbelief, then with resentment and animosity, they read such assertions as the "social construction of," the "fabrication of," the "invention of," the "archaeology of", applied to their science. In the social sciences the traditionalists have now absorbed this onslaught but in the sciences the attack may not be as successful if for no other reason than that the territory has a very stong barrier to entry. By the time one enters it and then learns how to manipulate science he is indoctrinated; there is no time for its deconstruction as problems, theoretical and practical that cover a wide spectrum, are thrown on his lap. The critique has had little impact because of the astounding successes especially in utilitarian items of technological content. But now science has reached its own frontiers and recognizes the deep questions. These are now becoming the subject of intense discussions.

My intention is not to provide an exegetical or historical view of physics. Rather it is to contribute a supplementary point of view that addresses a longstanding and deep problem in physics: namely, the process of measurement and experiment and, within the mainstream of physics, their relation to the theoretical constructs. I recognize that the physical world at this stage of our understanding can best be described in a quantum formalism. Yet, this is not the way the macroscopic world is comprehended. When it comes to discussing measurement — the quintessential quantum problem — for the objectives of the monograph I nevertheless retain a hybrid outlook: a Cartesian dualism mixed with a severed von

Neumann "chain." While dualism and the separation of mind from body is attributed to Descartes the paternity of the mind/body split may be a later myth: "Long before we read him, the famous conjunctions — mind and body — are impressed on us as his"[16] and hence contemporary authors on the subject of "consciousness" that make much of the Cartesian theater[17] should, perhaps, qualify their statements; apparently Descartes (1596-1650) was subtler than that. According to Reiss, he used this mind/body cleavage in much the same way as suggested here (a "passage technique," an interim strategy enabling him to get from the known to the new) that is, something akin to a working theory as we might say today. Similarly linguists, such as Chomsky,[18] assign language to "deep structures" in the brain without worrying about their historicity and evolution. In a sense, by cleaving history at some point, linguists avoid the vexing and difficult issues of origins — at least this is my explanation of their attitude. On the other hand, language may have snapped and expressed itself in a burst — a phase transition-like event (as in magnetism, superconductivity, and superfluidity), a symmetry breaking operation, a state change — so abrupt that evolutionary thinking may be incapable of adequately addressing it. Characteristic in this respect is the exchange of views between an evolutionary biologist and a mathematician (platonist in the extreme) who proved unable to agree on a common platform of understanding as the

[16] Sylvana Tomaselli, "The First Person: Descartes, Locke and Mind-Body Dualism," *Hist. Sci.*, xxii, 1984: 185-205 and Timothy J. Reiss, "Denying the Body? Memory and the Dilemmas of History in Descartes," *Jour. Hist. Ideas*, **57**, 4, 1996: 587-607.

[17] Apparently the evocative designation "Cartesian theater" was introduced by Daniel C. Dennett; however, the metaphor was used in Descartes' times. Indeed it is conjectured that *Le Théâtre Anatomique* of Caspar Bauhin furnished the description of Descartes' "glande H" — the pineal gland.

[18] Noam Chomsky, *Aspects of the Theory of Syntax,* MIT Press 1965.

opinions were quite divergent.[19] My work is *formally* along the Cartesian dualism lines arguing that the objective of those involved with consciousness and the such is to show (better yet, prove) how to get from physics and biology (matter, the Cartesian *res extensa*) to the sensation and clarity of mind (the Cartesian *res cogitans*) and explain the reason that duality is so intuitively functional. In the interim, and until such time as the puzzle of consciousness is resolved, I posit the "virtual Cartesian theater." In a sense I am using the same "passage techniques" to circumvent a difficult problem. This is a pragmatist's position that permits inventive activity, a process which could be thwarted by deeper inquiries. More detail on these caveats is furnished in the appropriate sections of the manuscript. I argue that the idea of *value* must be incorporated in the substance of thermodynamics and in the meaning of measurement and information; this necessitated some excursions and explorations into the territory of economics, biology, and thermodynamics.

The research and the writing of this essay progressed without support from any institution or grant; none was asked and no disclaimers are, therefore, needed.[20] The gestation and, then, the writing took more time than I care to remind myself. It was a conscious effort towards an interdisciplinary interchange of ideas and concepts, thus challenging the gospel as written by Schumpeter who "declared the independence of theoretical economics from psychology, sociol-

[19] Jean-Pierre Changeux and Alain Connes, *Conversations on Mind, Matter, and Mathematics,* edited and translated by M.B. Debevoise, Princeton University Press, 1995.

[20] This should be qualified as follows: in 1973-4 I had received a grant from Exxon Corporation for research on energy policy. It was then that one section of Chapter II was researched and written. I would like to thank Drs. J. Hansen and S. Stamas for the support.

ogy, biology and ethnology. Economics — Alfred Marshall to the contrary — have nothing to learn from biology, nor from ethnology."[21] Big words still echoing in the corridors of departments of economics and other fields. More recently Ernst Mayr[22] declared the "autonomy of biology" from physics and its ostensibly reductionist approaches. The territories of the trans-, the inter- and the intra- are rapidly being declared off-bounds. Yet, biology has learned from economics,[23] a discipline further up the hierarchical ladder. The interest focuses in the approaches to behavioural expression in the two levels of functioning. As I show in this monograph, the learning continues.

Many thanks go to Professsor Robert S. Cohen who read and patiently critiqued the text and to my wife Professor A. Lily Macrakis who, helplessly, saw the various editions follow one another.

Belmont, MA 1993-96

---

[21] See ftn. 3.

[22] Ernst Mayr, "How does biology differ from the physical sciences?" in *Evolution at a Crossroads: The New Biology and the New Philosophy of Science*, D.J. Depew and B.H. Weber, eds., MIT Press 1985: 43-63.

[23] For a review of the current thinking on contacts between economics and biology see J. Hirschleifer, "The Expanding Domain," *The Am. Econ. Rev.*, **75**, 6, 1985:85. Also, J.M. Smith, *Evolution and the Theory of Games,* Cambridge University Press, 1983. What remains securely common among many disciplines is mathematics. Economists have made extensive and fruitful applications of mathematics in their field and biologists are following on similar paths: see T. Fagerström, *et al.*, "Biologists Put on Mathematical Glasses," *Science*, **274**, 20 December 1996: 2040-1.

CHAPTER II

ECONOMICS

Le néant et l'être[1]

THIS CHAPTER IS A SORT OF A MEMO TO MYSELF. It advances no new theories or techniques. In it I attempt to review the question of capital in order to ensure (a) that there are no misconceptions as to parts of the theory which I subsequently use and, (b) that it is clear that the objectives of economists and mine are not the same. The reason is that invariably economists faced with the issue of origins of capital shrug their shoulders: it is not a subject in the mainstream of economic research and thinking as we shall presently see. At the same time physicists and biologists (less forcefully) will also claim that they do not understand the theory of capital; for them the theory is simply irrelevant. I

---

[1] Inverting the order of the existential question poses the fundamental question of physics and biology. Were we to assume that nothingness (néant) is the negation of being as well as the collection of particles out of which the assembly of physical and biological systems can be initiated and completed then, we have a good summary of the subject of my inquiry. It could be complemented with the sequence Néant → Chance (Hazard) → Necessity → Being. These concepts require a new imagery. What is a néant? Matter unstructured? Presumably we *know* — in the sense of Vico — what is necessity. But what is being? A result of an original cosmological disequilibrium? See also Chapter IV the discussion on *vacuum* and the *plenum*.

will try to show in the following chapters that this should not be the case.

1. Capital Theory

The word *capital* is encountered daily but this should not allow anyone to believe that it is also a trivial notion. The professional quandary surrounding capital theory is well illustrated in the introductory chapters of Bliss' work (1975). However, our purpose here is quite different and cannot follow the labyrinthine route to wisdom. Suffice it to say that the subject is anything but settled and it may well be that the notion of capital is itself evolving, never remaining the same. The point of view adopted here is that the objective of Political Economy is to invent the future and negotiate with it. To the extent that analysis of the past contributes to the identification of some "invariants" it is welcome as an important source of positive economics. These invariants, it is suggested, refer to the physical capital and its accessories as well as to the concept of interest which may be interpreted as a measure of success in replication and of the limitations of the biotechnological capital. My intention here is not to prepare a text on the economics of capital or a critique of the systematics of capital theory; that would have been a foolhardy enterpise even if attempted by a specialist. Rather, my objective has been to search the economic and related literature for clues on the origin of capital and its relation to energy: I needed to get an "impression" of what seems to be the central theme of capital theory while looking for clues from the phenomenology of economics that may have some impact on the later stages of my inquiry.

In this section I briefly report on my findings. First, I have concluded that an expeditious (but certainly not without blemish) entry into Capital Theory is to consider the Hicksian two-inputs-one-output simplification of the neoclassical theory. In this model the *macroeconomic* production (or, transformation) function is described by the production function[2] $F(K,L;t) = Q$ where the economic output of goods and services Q, is assembled (produced) at time t, by employing K units of capital ("machine-hours") and L units of labor ("man-hours") as inputs. This is anything but a trivial representation. It is the workhorse of theoretical economics but it is never quite made clear what it represents. Is it an ensemble average over many possible reifications of the economic world? How can the variables K and L be associated with observable measures given by econometric data? For simplicity I do not introduce the customary details such as transfer payments, taxes, and all the items needed to make the function realistic. Stripped to essentials, $Q = C + I$; i.e., production is all committed to investment, I, and to consumption, C. I is the rate of change of capital so that

$$I = \dot{K} \equiv dK/dt.$$

[2] For an introduction of the theory of production functions see Johansen (1972); on their use in growth theory consult the texts of Intrilligator (1971) and Burmeister and Dobell (1979) and the very extensive literature. For its use in macroeconmics refer to professional publications prepared by Chase Econometrics, Data Resources (McGraw), Wharton, etc. For the foundations consult Koopmans (1957).

At this level of discource, I assume that the concept[3] of Capital is a familiar, intuitive, protoeconomic notion. In this manner I evade the historical route with its meanderings which are so easy to roadblock. The above representation describes and stresses the interplay of capital and labor in the production system and refers to collective, aggregate, economic phenomena; to complete the picture I must, in addition, accept that the economic system is driven by the utility of consumption and of leisure as the latter is moderated by the discount rate. This is the *bare bones* but abstract view of "neoclassical" economics or, better still, the canvas on which much of the complexity of economics can be woven. While the picture is not simplistic, it nevertheless glosses over important detail. For example production appears to be independent of resources — as if a *deus ex machina* provides all the necessary inputs — while, miraculously, assigning some price to these as a measure of their scarcity. However, with the concerns about energy inputs in the economy and the intuitive knowledge of the pervasiveness of its use, it was felt, early in the seventies that access to an analytical representation highlighting the energy and the material inputs into the economic accounts should be available. This can be accomplished quite naturally with the use of Input-Output (I-O) tables (see figure II-1) and the following chapter, where Energy (E) and Material (M) inputs and their transformations may be readily identified and, at least in principle, used to estimate the costs of alternative technological realizations or, virtualities, in the

[3] "concept" refers to an undefinable, such as distance in geometry. It is an *a priori* truth in the Kantian sense.

manufacturing sectors.[4] We will see below that the I-O tables contain explicit transactional information on all aspects of the production process just one level below that of F(K,L;t), and they are closer to the microeconomic perceptions of the economy because there is a possibility to correlate the concept of K with a machine and to connect energy with a specific resource.

	M	*E*	*C*	*I*	FD
(M) Materials	x_{11}	x_{12}	C_1	I_1	X_1
(E) Energy	x_{21}	x_{22}	C_2	I_2	X_2
(K) Capital	K_1	K_2			
(L) Labor	L_1	L_2			

where

$$F(K,L;t) = C + I = Q;$$
$$K_1 + K_2 = K; \quad L_1 + L_2 = L;$$
$$C_1 + C_2 = C; \quad I_1 + I_2 = I$$

Figure II-1 The structure of an Input-Output table.

An elementary (2×2) Input-Output Table with two sectors, with Final Demand (FD) and Inputs of K&L. A more sophisticated description would involve an I-O matrix with many more elements. The capital stock allocated to the processing of materials and for energy transformations is summed up to yield the collective capital K; similarly for L, C and, I. On this (summation) principle one transfers from microeconomics to macroeconomics.

[4] The literature on I-O tables is very extensive but for the original development consult Leontief (1951). His is a detailed analysis of supply under conditions of fixed technology.

For the immediate concerns of this essay I, therefore, adopt the position that the fundamental economic production function at the level of the I-O representation must be a function of four distinct imputs namely Capital K, Labor L, Energy E, and Materials M. K with L and, E with M, form natural pairs respectively; the first pair, we may call generic Capital; after all, physical labor, except for its empathy, *is* capital. Similarly generic materials, cover both the *inert* state and the *active* state of matter; the former is for structural use while the latter is energy-containing. These considerations naturally lead to a two- or, four-factor production function $F(K_g,M_g)$ and F(K,L,E,M) respectively which are the forms[5] fitting the needs here. Formally, and quite generally, we can say that generic K_g and M_g are assembled and consumed to produce a new piece of capital. The production function specifically implies transformations of energy to keep the capital in operation so that it can process materials to build more (physical) capital. All along then, I shall consider the K_g-M_g interactions and the accounting implied by the input-output table as essential in support of the ideas in this essay. Note now that to connect the I-O representation with the production function as such we must introduce the matrices for C and I as well as for K and L shown schematically in Fig. II-1 for a two sector economy.

The procedure discussed gives the appearance of graininess contrasting with the economic continuum implied by the smoothness of the production function; in its abstraction it is also quite distant from the individual actors (persons, classes). At this level of aggregation K becomes the surro-

[5] see Hudson and Jorgenson (1974).; they have introduced coefficients that are price sensitive and hence a methodology better fit to address policy analysis and incorporation of the I-O into models of econmic growth.

gate of the actual physical capital. It is still called capital but it is measured in allocative units (dollars say), which, unlike the physical units, are homogeneous but not necessarily permanent — reflecting as they are, values. This (aggregate) capital can be endowed with the ability for improvement while L remains the "unimprovable" biological capital, i.e. labor. The production function thus allows for possibilities of substitution of the *engineering* for biological capital in the large. The production function discerns between capital and labor as necessary inputs in production but this is what differentiates economics from biology: the latter represents capital and labor as consubstantial.[6] Whatever the underlying mechanisms in production, political economy focuses on the interplay of the two components of K, capital and labor. All other activities take second place in the engrossment of humans to determine the shares of income between themselves and capital as well as the substitutability of capital for themselves. Indeed, the Marxian analysis has given considerable weight to the relationship between labor and that component of labor (the proprietors) which uses human capital elastically connected to the physical capital (machines).

The economic world is characterized by a long-run flexibility, an *almost* universal elasticity of substitution and of demand and is anything but invariant;[7] it is inventing itself and, therefore, it is dialectical in the Hegelian sense, i.e. open-ended and indeterminate. *Positive* economic theory addresses the immutable regularities which, although applicable at any time and situation do not preclude invention

---

6 Ὁμοούσιον, to use a word coined during the Synod of Nicea (A.D. 325).
7 See Monod (1972) and his suggestion about "invariance."

and institutional differentiation, both unpredictable and novel, but deriving from the *normative* aspects of economics.

In the precapitalistic societies, biological capital consisting of men, animals and plants, would be the only type available.[8] Plausibly, the early differentiation between labor and capital in the environment of scarcity marks the beginning of human history. This idea is further eleborated in section 3 and Chapter III where I support the proposition that *scarcity* is the preponderant constraint and the dialectic of economics. In its most elemental appearance it is scarcity of materials and of energy as well as of the time required for the assembly of capital from energy and material inputs to which we refer; to this we must add the scarcity of information while deciding where and how to invest capital. This concern must have been in the mind of Hayek who chides the "engineers" for their one-dimensionality in thinking about thermodynamics and the second law in connection with resource economization:[9]

---

[8] The question is not novel. Fischer (1930) refers to the "psychic returns of the *human machine,* while Charles Turgeon asks whether "L' homme est un capital?" and he answers "Non, ou plus exactement il l'a été et il ne l'est plus;" see *Révue d'Économie Politique*, 38e Année, no1, 1924: 65 and 73 for respective quotations. More generally Paul Levy-Beaulieu in his *Traité théorique et pratique d'économie politique*, (Paris 1913: T. I, VI) was writing that "Quand on examine les choses de haut, on voit que toute civilization materielle est un merveilleux phénomène d'incessante capitalization materielle, intellectuelle et morale."

[9] Hayek (1955: 96, Chapter 10). Thermodynamically, the possibilities exist within some rather large limits but it is the discount factor that comes into the picture quite forcefully. An alternative way of stating the same idea is the following: although substitution of capital and labour for expensive fuel is consistent with the basic tenets of economic theory and practice, the question nevertheless remains "whether the isolated allocation of capital to conserve energy can be extended to all processes in the economy that require energy inputs without affecting the growth patterns: an engineer's design possibility frontiers are always moderated by macroeconomic constraints," (Macrakis, 1973). Capital is indeed scarce because it is developed by accumulation in time.

> ...The engineer's 'technical optimum' proves frequently to be simply that method which it would have been desirable to adopt if the supply were unlimited, or the rate of interest were zero, which would indeed be a position in which we would aim at the highest possible rate of transformation of current input into current output. but to treat this as an immediate goal is to forget that such a state can be reached only by diverting for a long time resources which are wanted to serve current needs for the production of equipment. In other words, the engineer's ideal is based on the disregard of the most fundamental fact which determines our position here and now, the scarcity of capital.

This is an extraordinarily clear description of an economic problem and it reveals an important mechanism that connects the collective with the isolated decision maker. I will elaborate on this point later; for the moment I note a dichotomy: at the engineering design level, there is an apparent freedom in the selection of the process that will economize in one or the other of the input leading to trade-offs (substitutions) respectively between capital and labor and, capital and energy. Yet, when all microevents are summed up and then analyzed in their aggregate historical performance as revealed with econometric time series, say, some data indicate that in the large, the manufacturing sector is characterized by complementarity between capital and energy inputs. This counterintuitive result was unexpected since in economics substitution prevails. As a consequence, the making of policy rests on a perception approaching the

---

Unlike natural resources which, even if exhaustible, can be tapped at will — ...until dry — capital is a temporally integrated quantity which cannot be suddenly switched on or changed.

status of a fundamental dogma, namely, that of substitution. In the long run this is apparently the case. However, for the short run and the short run may be long indeed (months to eons) phenomena of complementarity can be important. The econometrics of the Jorgenson school seem to indicate that indeed, and for at least the *manufacturing sector*, and in the short run, complemetarity prevails; it is a hotly contested finding. A way to interpret this phenomenon is to look into the design and construction of equipment by all manufacturers together as they attempt to conserve on one resource, in this case energy. The resulting run on capital, a scarce resourse in intertemporal terms, will make it expensive. Not every one can conserve by just substitution although many can do so by invention and change of technique.

This far I have dealt with a quasistatic picture of the economy and of the production function which images it. With the dimension of time included, however, the "reference" description of the macroeconomic process is given by the *reversible* (i.e. retrodictive, when $t \rightarrow -t$) equation

$$F(K,L;t) = Q = \dot{K} + dK + C \qquad (2)$$

Here I have allowed for the replacement of a fraction δ of the original capital in order to make up for losses resulting from wear and tear. K itself may be considered to be in the nature of a fluctuating term which could even cancel out δK; not all experiments succeed in advancing capital formation. The fluctuating capital allows us to escape from the "essentialist" approach criticized by Blaug (1984: 124 and *passim*), the "essence" of the economic process being reproduction of capital. Note, however, that in the Marxian world it is reproduction of the power structure while, in the sense of Ricardo and Sraffa, it may also be the reproduction of physical capital (the corn-seed cycle).

Koopmans (1973) has called this autocatalytic reproduction equation, the *economic equation of motion*.[10] The dynamics of this equation, under some restrictions of no importance to the present argument, can be reduced to a one-dimensional process by using intensive variables normalized to the labor inputs. By writing

$k(t) \equiv K(t)/L(t)$ for the capital - labour ratio

$c(t) \equiv C(t)/L(t)$ for the consumption per head

$f(k) \equiv F(K,L)/L = F(K/L,1) = Q/L$ for the per capital gross output,

we obtain

$$f(k) = \dot{k} + \delta k + c \qquad (3)$$

for the equation of motion with δ, as before, the rate of capital depreciation. In this form, it has become the subject of a voluminous mathematical literature often with tenuous relation to day-to-day economics. As used, the economic equation of motion is a phenomenological one of reproduction and it describes economic phenomena at the collective level. At this level the concept of depreciation loses all its

---

[10] Is it legitimate to call this the "equation of motion"? The notion of equation of motion derives from Newtonian or Hamiltonian Mechanics. Perhaps it is referred to in economics as "equation of motion" with tongue in cheek. In fact it has little conceptual connection with the physical equations of motion despite the fact that it too can be derived from a minimum principle. The reason is that it reflects an accounting scheme rather than a physical law. It is more akin to the equations used to *control* a spaceship.

physical content and becomes an accounting measure describing a continuous process of decay to be made up by additions to capital. By contrast, at the microlevel, even including whole sectors such as the electric utilities, the depreciation takes on Holmesian — "one hoss shay" — characteristics.[11] The equation of motion by itself has no explanatory value: it does not tell us how the first capital was formed, why the capital decays, what is the purpose of consumption C or of leisure and of capital formation. Apparently these *concepts* must be self-obvious although they have no relationship whatsoever to the concepts or tools used for the description of nature. Yet the equation of motion would amount to little without a driving element (force) which, in economics, is taken to be the utility of consumption $U(C)$. It was F.P. Ramsey who in his 1928 classical paper on "A Mathematical Theory of Saving,"[12] introduced the idea of optimizing the time-integrated utility of consumption as a negotiated interrogation of the future. In general, one wishes to maximize

$$U(\varrho) = \int_0^\infty e^{-\varrho t} U(C)dt; \qquad \varrho \geq 0 \tag{4}$$

Here C is the flow of consumption at time t, while ϱ is the discounting rate for the future utility flow $U(C)$; Ramsey thought that in fairness to future generations, ϱ should be zero . Expressions 3 and 4 form the backbone of all theoretical and practical work in the field of economic growth. As

---

[11] W. Nordhaus has used the "One Hoss Shay" model in the description of depreciation in the economic equations: the Holmesian machine works perfectly well for 30 or so years and then it just disappears. δ is a phenomenological "evaporation" coefficient that reflects microeconomic behaviour and accounting practices.

[12] See *Econ. J.*, **38**, 1928: 548-550.

Koopmans (1974) had suggested the theory of "optimal economic growth is the economist's counterpart to the engineer's optimal control theory." Indeed, the economic intertemporal maximizations are no different in mathematical content than the problems of missile trajectory formation under constraints (to minimize fuel, say). But this is also their flaw: as with missiles, the alternative economic paths are pretty much preordained by initial conditions and inelastic in this representation. On the other hand, in the real world economic paths are quite open, at least for the long run; errors are not collectively or invariably fatal. In any case, the collective (normative) decisional problem of a society is to select an ϱ. In poor democratic societies the discounting is high and as R.F. Harrod wrote in 1948 the pure time preference (= discounting) of future utility is "a polite expression for rapacity and the conquest of reason by passion."[13] In richer, democratic societies, the discounting may be lower and in societies with central planning and heavy capital investment programs, ϱ may still be lower, favouring the future, yet unborn, generations. On the other hand $\varrho = 0$ presupposes a perfectly clear, uninteresting, future. Because of uncertainties and the stochastic nature of capital reproduction, ϱ must have a finite value.

Avoiding the philosophical and ethical questions concerning inter- and intra-generational equity issues[14] one may

---

[13] *Towards a Dynamic Economics,* Macmillan 1948.

[14] For which see J. Rawls, *A Theory of Justice,* Harvard University Press, 1974: "..it is a natural fact that generations are spread out in time and actual exchanges can take place between them in only one direction. We can do something for posterity but it can do nothing for us" (pp. 284-293). As we shall see in the following section, Sartre expresses a related idea by stating that interpretation starts from the future!

stay with the more practical and suggestive view of Keynes about the nature of capital:

> It is by reason of the existence of durable equipment [i.e. physical capital] that the economic future is linked to the present.[15]

With a rather condensed synopsis on the "theory of capital" behind, I add some remarks which may be useful later:

First I note that as Dorfman pointed out "capital (stock) *change* is a function [f(k,x,t)] of its present standing."[16] The same, of course, applies to the quantum mechanical function ψ: $d\psi/dt = H\psi$ where H is the Hamiltonian; its change too is a function of its present standing. The implication is that there is no action at a distance; the equivalent economic implication is that you cannot increase capital where none exists. These results relate to the fact that the Huygens' principle (in propagation) and Fermat's theorem (minimum paths) are two sides of the same coin and equally valid in economics if the proper analogies are kept.

Second, I conjecture that the phenomenological equation of motion refers to an iterative, interactive process which is

---

[15] *General Theory,* p. 146.

[16] R. Dorfman, "An Economic Interpretation of Optimal Control Theory," *Am. Ec. Rev.*, **59**, 1969: 817-831. In this essay Dorfman provides a well articulated economist's approach to capital theory while giving a tutorial on the application of optimal control *theory* to economics. (The economics of optimal control is just another matter!) Dorfman's world is in a space where the *value* of capital counts; not its physical structure; if the idea works it is because so much is placed outside the system, exogenously. The tacit assumption is made that the capital system is in place by some magic while aleatory issues are only insinuated. How decisions are made, how the signals of prices circulate are of no concern; they are subsumed under the idea of "profit."

the equivalent of the *quasi-particle* in the many-body theories in physics. This is a "particle" which behaves like a wave or a distinct particle depending on the nature of the measuring stick. The economy as a whole represents the collective behaviour of all its members while the individual actors perceive the totality through their interaction with the quasiparticle. Recall that Keynes had taken the first step in this direction with his bold assertion about the functional relationship between aggregate consumer expenditures and national income. Indeed, breaking away from the methodological constraints of the Smithian view of the world which attributed all economic phenomena to the manifestations of individual behaviour is the revolutionary contribution of Keynes to economics. This model of an economic system becomes intelligible if one studies the interaction between the two levels of representation of the process: note that the intertemporal calculations at the collective level are done within the physical constraints of available total capital and labor inputs. However, should the planning demand different K-L inputs, or should the constraints on resources demand variations, substitutions, or new technology at the KLEM level, then, the coefficients $a_{ij}(p)$ of the I-O table must vary (if at all possible with the price level, p) to reflect and accommodate the new facts. With this flexible representation one may imagine the real life interconnections between the collective and the microeconomic picture; the latter is closer to the decision level of the economic actors (be they investors in the neokeynesian scheme of things or struggling classes in the neomarxist view) that react to the collective signals (cost of money, say) for the adaptation of production, investment and, consumption patterns to new realities. The interaction is taking place in the economic *field* (to use a term from field theory in physics) which connects the collective level of the economy with the single actor. The two economic levels are quite distinct and are interconnected through the (information?) field, the medium in which signals and perceptions pass from one level to the other. The mechanics of the signal paths cannot be captured with formulae, although, to a degree, econometrics attempts to conjecture and identify general rules pertaining to this interaction mechanism. However complex, the interaction of the individual actor with the collective *quasiactor* remains at the root of the difficulties in the study of economics. A deeper understanding of this interaction can be expected to have a fundamental political and policy implication. It is also plausible to expect that this collective-independent particle interaction may have its counterpart in the early developmental stages of the elemental biological replicative mechanism whereupon the totality of a reproduced set may have had a restraining influence on the molecules about to be synthesized (see following chapter).

This section has been written in the spirit of the neoclassical schools of economics; without making it explicit, I have followed rather closely the mapping of the economy in an intertemporal, utility maximizing, wealth targeted, neoclassical model of the economy which allows for resource constrains through the variable coefficients of the I-O table. The objective, however, has been to underscore the function of the reified capital as the means for intertemporal, intergenerational communications. This also brings the section to a conclusion.

How can we know the dancer from the dance?[17]

2. Whence Capital

In the first section I have referred to *capital* and *discount* as primitive concepts, not as quantities subject to a definition. This permitted the setting of an "equation of motion" and the introduction of a formal intertemporal calculus. But for our purposes here we must go deeper and ask the question *Whence Capital?* What is the origin of capital and of the discount rate?

In searching the literature of economics one cannot find a definite statement about what is capital and what are its origins. On "what it is" there is abundant discussion, much disagreement and, surely, many misconceptions. Böhm-Bawerk,[18] for instance, said that "there are three theories current concerning the formation of capital. One ascribes the origin of capital to saving, another to production, the third to both factors." To the present time the ambiguity has not been dispelled. Because of the many faces that savings and interest take, it is difficult to assign them to a specific and uniform type of *capital.* While terminology and the origin of words is functionally sterile, we may note that the origin of the word unambiguously implies that it refers to animals: head-counting. That *is* the origin of the *description*

[17] W.B. Yeates, 1928.

[18] 1959: Chapter V, "The Theory of the Formation of Capital." In addition one may read with profit the remarks on the "State of Capital Theory in Keynes' time," in Leijonhufvud (1968: 212) and Brown, M. *et al.* eds., *Essays in Modern Capital Theory,* N. Holland, 1976 where the difficulties of pinning down "capital" are discussed. See also Kregel (1976).

of capital. Introductory courses on economics on the other hand use the metaphor of *seed corn* — as an instrument for reproduction and consumption — to illustrate the meaning of capital. Thus the fundamental characteristic of capital must be its durability which is extended through the mechanism of reproduction. My searches have yet to identify an effort addressing the physical and abstract mapping of capital which, in its modern context, will incorporate capital-like, capital-mimicking or, capital-supporting and maintaining operations; that is, indirect activities beyond those of reproduction and accumulation which, in practice, also command interest — the measure of discounting the future.

In the observation and study of the economic scene one encounters too many components of the economy that — chameleon-like — take the appearance of capital in allocative terms. Indeed, only a few sectors in the economy retain any semblance to the physical (real) capital idea and these can be located within the manufacturing sector which does *reproduce,* as it were, its own machines for additions and replacements along with the processing of inputs which end up as consumption goods. Fundamentally, by ***physical capital*** we refer to the machines that process goods and produce other machines. All other items such as buildings, inventories, roads, land, savings, etc. are the peripherals to capital which are nevertheless assigned capital-mimicking valuations (measures) in accounting practices. An example of (physical) capital which is nearly homogeneous and imperceptibly (if at all) improvable is the human body; yet exactly because of this characteristic it has been differentiated and called *labour*. Marx described it as *living labor* to separate it from *dead labour* — the machines.

So great is the frustration encountered in attempts to describe the "essence" of capital that Bliss (1975: 7) suggested that "it is best not to worry about that, because capital is many things to different men." Perhaps, then, the way out is to remain pragmatic and feign that we understand the issues. The reason for these difficulties originates from the mechanism of information exchange between the collective, societal decisions on capital formation and the allocations between savings and consumption at the micro-level. These mechanisms force the perception of capital-like procedures in most of the economic activities whether they are related to the making and maturation of wines, the planting of vineyards or, the factories for the manufacture of potatoes or silicon chips. The collective negotiation with the future is performed through the formalization of interest rates or the rate of return on capital.

Solow (1963) in his *Capital and Rate of Return*, expresses his bafflement as he searches for a comprehensive definition of capital and concludes (in despair?) by adopting a common denominator of what, in practice, is operationally accounted for as capital; he remarks that the "central point in capital theory *should* [my italics] be the rate of return on investment" (p.16). This, he adds, is the "middlebrow" capital theory. The "highbrow" version is given by the intertemporal concepts discussed in the previous section; this theory looks too much of a theoretical construct and an abstraction to the decision maker who, as a "particle" in the collective economic sea, prefers to use what he observes in practice in the form of interest rates and capital accounting valuations rather than to search for deeper meanings. The highbrow theory on the other hand rests on the concept of the discount rate, a measure of which can be extracted from econometric data but for which there is no procedure to develop out of "first principles."[19]

But what then is the origin of capital, of utility, and of the discount rate? When, how and in what form did these *observables* make their appearance? How did the economic process ever begin? These are the fundamental scientific questions, seemingly of no interest to the actuality and ephemerality of economics. The question "whence comes the first capital?" has been asked before, however. Marx[20] suggested that capital did originate from violence and the will to power and that the incentive for the primitive accumulation of capital had no other origin. To Marx, capital was not the physical stock but the "strategic financial element." Marx and economists of all schools refer only in passing to the physical structure of "capital." Marx, neomarxists, and neokeynesians are all quite concerned instead with the techniques and the institutions which allowed (and still allow) for control of investment, the attendant practices — or, Keynes' "animal spirits" — to secure *it* and its returns, and the psychic (alienating) impact of the process on society and on labor in particular.[21] In this sense capital "holds the key to the distribution of income," as

---

[19] But consult a cogent analysis and determination of the collective discount rate by R. Dorfman, "An estimate of the Social Rate of Discount," Discussion Paper # 442, 1975, Harvard Institute for Economic Research, Harvard University.
[20] K. Marx, *Capital*, English transl. Moore, S. and E. Aveling, New York 1936: 784-785.
[21] This summarizes extensive and politically important ground. More technically, I imply distributional issues and property rights (in this connection consult Kolm, 1983), the most important themes in normative and positive economics; I also imply the dynamics of the formation of firms, a subject of some import to my subsequent discussion (in this connection consult Steindl (1965), etc. In short ... all of economics!

Bliss (1975: 11) righfully asserts. To all this one must add Gershenkron's views which attribute the "spurt of industrialization" to a prespurt (accidental, unfocused, unintentional?) accumulation of capital which led to a "beginning."[22] This is the only meaningful question, he argues, while he derides the search for the "first capital" thus predesposing (and providing ready–made ammunition to) other economists. By contrast, "capital" in this essay is taken to be a universal concept originating independently of distributive, violence, and decisional issues.

All modern industrial states are "capitalist" in nature in the sense of consciously developing physical (technological) capital for production as an intertemporal objective in economic growth. But the operation of a capitalist market system may be seen from three different theoretical perspectives, the neoclassical, the neo-Marxian and, the neo-Keynesian.[23] The last two are closer in perspective in that both assign predominant functions to the capitalist class. In both, the saving propensity of the capitalist class is the key driving force of the system but with the following differentiation: a "subsistence" wage is characteristic of the neo-

---

[22] Gerschenkron (1962: 33 and 99); his spirited critique quotes from Marx and Keynes. What he wants to show is that at some point, out of random capital-like events some instability in the system ocurred which, not unlike the conjectured "punctuated equilibria" in biological evolution (see Chapter III), triggered (clicked) a conscious accumulation of capital out of profits. Capital existed before but it had not been marshalled collectively for a purpose: to produce a new economic "state" — capitalism. See also p. 9 of the Introduction.

[23] The reader must complement this effort with the post-Keynesian and the neomarxist (structuralistic) critique of production for which see Marglin (1984), for a lucid classification and analysis. This is a controversial monograph but contains a rather unique attempt to provide a much needed coherence in the field of economics. The appropriate criticism leveled is that Marglin's works are "exercises in optative economics" always implying a "vision of what might have been ... accentuating the negative." (D.S. Landes, "What Bosses Really Do?" *J. Econ. Hist.* XLVI,3, Sept. 1986: 585). Caveat lector.

Marxian view while investment demand characterizes the neo-Keynesian view. The neo-Marxian view separates the "capitalist" class from the working class (labor): intertemporal decisions are made by the former. By contrast, the neoclassical and the neo-Keynesian view is that the "household," through its utilitarian preferences and life-cycle saving characteristics, is the center for decisions — at least indirectly. The various views and their hybrids are providing different economic models for the workings of a capitalist economy. It seems clear that decisions affecting capital accumulation, are not made at the "household" level.

Modern marginalist theory has accomplished a significant step towards the separation of the fundamental properties of the universal concept of capital and its wages from the emotionalism, scarcity, the legal structure[24] and conventions and the ethics of income derived by capitalists. Capital in this view is disembodied and appears as an abstraction having only remote relation to "head." Things are valued as having some price in both normative and positive terms and, decision makers — wherever they may be — decide how to physicalize (embody, reify) a virtuality; they are thus given the "promise" that in time, they will be permitted by the collective and within its legal structure and conventions, enough labour hours and materials to physicalize the capital, so that its wages acquire a formulation devoid of alienation. Before its reification, capital can be referred to as *virtual capital.*[25] Eschewing the dialectics of

---

[24] "It is only under the shelter of the civil magistrate," wrote Adam Smith, "that the owner of that valuable property, which is acquired by the labor of many years, or perhaps of many successive generations, can sleep a single night in security."

[25] The *virtual* is of course a mental state; it can be reversed in time, changed in no time, it has no localization. The moment of reification or better yet the mo-

economics (and the gaps between normative prescriptions and the actuality of strong inequities), I seek the universals of economics. The stylized description of economics adopted here is an abstraction in the midst of nowhere. It does not allow for a beginning (how did it all start?) and it does not provide a hint or a clue as to the mechanics of its operation. In reality, however, the economic system is a durable, albeit wear-prone and destructible, engine of reproduction; the equation of motion describes an invention of biological origin which has evolved in time and is now being experienced in the large through economic activity. It is an equation which is law-like, yet seemingly, not explainable on the basis of underlying physical mechanisms and concepts. It is a phenomenon. The biological and the derivative economic systems operate within the laws of physics (law-abiding); but they are innovations the laws do not contain. The initial conditions and the fluctuations of the economic equation of motion pose the same kind of questions as the evolution of life itself.

In the "precapitalistic" world i.e. before the conscious process of accumulation, the average capital may well have been zero: $\langle K \rangle = 0$. Fluctuations ΔK, may still be observed as random attempts to form capital structures; however, the mechanisms for reaching a state with $\langle K \rangle \neq 0$ or a finite integral of $\langle K \rangle$ over time (i.e. wealth) are not at all obvious. We are not only reminded of the "nihilated-nihilating" process in Sartre's *L'Être et le Néant,* but also of the process described by Walras as *par tâtonnement* which consists of "blindly feeling its way, since no one in the actual world is

ment the initiation of reification begins then an irreversible condition prevails. The concept has a parallel in the theory of measurement and the "collapse of the wavefunction" about which in Chapter IVB.

presumed to know in advance the parameters of the solution of the equations."[26]

3. *La Raison Dialectique*

In the attempt this far to recognize or develop a theory on the origin of capital it appears that we have reached an impasse. Economists circumvent this by cutting into some point in history, by assuming the existence of "capital" at the cut, and by looking into the future. This decoupling is the key element in the foundations of economics which thus liberates the theoretician and the practitioner from reference to the physical world and the debilitating reductionist argument. This decoupling permits the development of "predictive" tools which, however, appear to be extrapolatory rather than really predictive. In fact all econometric models are set with initial conditions and then run into the future without concern for the aleatory changes that might appear on the way.

My concerns, however, are of a different nature; for thisreason I have sought to draw on perceptions from outside the fields of economics and physics in order to develop a bearing for the subsequent steps and chapters. In the search of the literature for the origins of capital I have come across some striking philosophical descriptions — gropings might be more apt — which have some relevance here. In particular I refer to the work of Sartre (1960) which

---

[26] See William Jaffé, "Walras's Theory of Tâtonnement: A Critique of Recent Interpretations," *J. Pol. Econ.*, 75, 1967: 1-19.

in many instances appears too abstract, repetitive and convoluted, but which acquires some significance for the questions posed here. In his celebrated — but largely unread — *Critique de la Raison Dialectique* (CRD, for short), he writes:

> With the organism, dialectical time has entered into being since living can persist only by its own renewal; the *temporal* relation of the future to the past through the present is nothing else but the fundamental relation of totality to itself: it is its own future beyond a present of reintegrated disintegration (CRD: 167-168).

The last two words, "reintegrated disintegration" describe the idea of breaking away from the noise into the coherent capital formation. An interpretation is that an 'organism' and its reproduction is a manifestation of what I have come to designate as generalized capital, while 'totalization' implies the actual (mathematical) integration of capital over time which, net of 'evaporations' and 'depreciations' is by definition Wealth (W) in economics:

$$\int K_n dt \equiv W; \quad K \equiv K_n^0 + \delta K$$

where δ is a coefficient. "L'Histoire humaine est en effet, orientation vers l'avenir et conservation totalisante du passé,..." (CRD: 200). Indeed wealth (W) in economics is defined as the time-integral of capital, net of its depreciation. "Reintegrated disintegration" is just the economic process which consists of the evaporation of capital (in the terminology of economics) through wear in use and obsolescence, while new net additions to capital, often embodying new technology, coherently take place but with both processes randomly occuring. Equally striking is the following paragraph:

> Donc l'homme se définit par son projet. Cet être matériel dépasse perpétuellement la condition qui lui est faite;... C'est ce que nous nommons l'existence et par lá, nous n'entendons pas une substence stable qui se repose en elle-même mais un déséquilibre perpétuel, ...[27]

In statistical physics and thermodynamics, the non-equilibrium processes have been the subject of intensive research. In economics, on the other hand, the question of non-equilibrium has yet to be considered at all for it defies a theory with predictive or, even, heuristic power. The "equation of motion" of the economic experiment describes the necessary paths of capital accumulation smoothed over time. It does not describe the microdisequilibria that are necessary albeit not sufficient a) to initiate the economic process and b) to take an economic process away from the routine of recurrence (reinitialization) and into the path of history-praxis (practico-inert in the terminology of the CDR). The limited discussion by economists of non-equilibrium states explains also, the antidialectical procedures followed in the economic analysis in general. A possible exception is the work of Schumpeter that articulates a dialectical process as

---

[27] Sartre, CDR, *Méthode,* p. 95 which translates as follows: "Man, therefore, defines himself through his project. This material being, perpetually surpasses the conditions in which it finds itself...This is what we call existence and by this we do not imply a stable substance which rests on itself [on its earlier accomplishments], but rather a perpetual disequilibrium."

a mechanism for business cycles as well as the work of Gershenkron which addresses economic development.[28]

In economics, as in engineering practice, there is the element of improvement and of inventing something that is not there. It is not a matter of stochastic events or processes as encountered in physical problems, but, rather, of stochastic events that latch on to something new. Thus the essence of economics does not lie in the descriptions of, say, fully efficient markets, but in the non-equilibria which allow novelty by composition. The economic agent must be in a state of non-equilibrium, otherwise his functions would be meaningless, as the system would only be a clockwork, a perpetual uninteresting input-output activity[29] Indeed, "for a world that is always in equilibrium there is no difference between the future and the past, there is no history."[30]

The driving source in the economic process is expressed in the form of an intertemporal maximizing principle referring to the "utility of consumption" $U(C)$, discounting the future at some finite interest rate. "Interpretation starts from the future," so that man "is a flight out of the present into the future."[31] This description is the essence of the process because it, implicitely, relates to the behaviour of economic actors in the face of actual and of perceived *scarcity*. Without providing a physical or philosophical reason for the source of scarcity, the phenomenon of Utility, perhaps,

---

28 Schumpeter (1949: 84); Gershenkron, *Backwardness, loc. cit.*
29 Inflation, for instance, may be just one of the observables of this process when readjustments occur in, say resource utilization and substitution or, when innovation lags in the face of persistent demand and timid public employment (or distributional) policies.
30 Robinson (V, 1980: 173).
31 Sartre (1497: 469).

summarizes succintly the response to scarcity of the historical, or dialectical, being. By dialectical one refers specifically to the being that negotiates with the future in order to assess *scarcity*, calculate its effects and then proceeed in obviating it through praxis, invention, novelty, migration and, even, aggression.

Sartre gives great weight to scarcity: "scarcity defines a particular relation of the individual to his environment" (CRD: 20). While the discipline of economics addresses the problem of allocations in the field of scarcity, as part of the economic activity, it does not axiomatically consider the attempt to obviate scarcity as the driving force for the economic *diergodicities*[32] (savings, the life cycle mechanisms, etc.) In a significant and commanding phrase, Sartre remarks that

> l'Histoire nait d'un déséquilibre brusque qui fissure a tous les niveaux la societé; la rareté fonde la possibilité de l'histoire humaine et seulement sa possibilité en ce sens qu'elle peut être vécue (par adaptation interne des organismes) entre certaines limites comme un équilibre" (CRD: 203).

Indeed, one could conceive of a world without History but that world would be one

[32] A "process" implies a stepwise change while with "diergodicity" I attempt to convey the transactional negotiating process. Not to confuse with "ergodic" which in statistics defines a process that, given time (often very long time) a system will return to the *status quo semi-ante.*

> où végéteraient des groupes humains demeurés au cycle de la repétition...[33]

While the importance of the non-equilibrium is recognized, the elevation of *scarcity* to a driving force essential in the making of history is significant. After all, economic descriptions of the intertemporal optimum exhaustion of exhaustible natural resources, elasticity of substitution and of complementarity are all dependent on the theme of scarcity.[34] In the spirit of the above discussion relating to the economics of resources, their scarcity and their exhaustion, I formulate the following four axioms which may be applicable to the post-Hesiodic economics (chapter I, p. 10):

---

[33] These passages respectively translate as follows: "History is born out of a disequilibrium which fissures (cleaves) society at all levels; scarcity founds (establishes) the possibility for human history and only its possibility in the sense that it [history] can be lived (through internal adaptation of the organisms), within certain limits, in the form of an equilibrium," and that one could conceive of a world without History "where human groups would vegetate remaining in a cycle of repetition."

[34] Studying the early growth models in economic theory we notice their insensitivity to resource availability and their unrealistic growth patterns which are amplificative rather than dialectic. The more recent models address scarcity based on estimates of resources and technology. Note, however, that even in the description of complex time-dependent physical problems one is continuously faced with the necessity to introduce new Hamiltonians to account for the 'before and after' realities in an experiment (see Chapter V on experiments). In an non-dialectic economic system the repetition implies a frozen technology (however reached) and therefore, a fully complementary relation of the factors of production. Change implies the possibility of substitution between factors of production at least within some limits. The question of substitution versus complementarity, however, is quite controversial in the relation of energy and capital in manufacturing; this controversy is central to the making of an energy policy where structuralist constraints may negate strong assumptions of elasticity inherent in neoclassical models.

E1 Scarcity of resources is the driving force of economic and historic diergodicities.
(The pure Marxist theory refers to the harsh exploitation of man by man as the driving force of capitalism. The neo-marxist theory refers to the class struggle while the neo-keynesian to the investment dynamics.[35] Whatever the picture, the driving force is the response of society to scarcity.)

E2 The estimation of scarcity is done through the discount rate in connection with the maximization of ophelimity under conditions of uncertainty:

$$max\ of\ \Omega(\varrho) = \int e^{-\varrho t}\,\Omega(C)dt;\ \varrho \geq 0.$$

E3 The *Equation of Motion* of economics is that of capital accumulation and reproduction in the field of scarcity:

$$A_S(t)F(K,L,t) = \dot{K} + C; \quad K = \langle K_f \rangle + \delta K_f$$

(Note, however, the reversibility of the equation — a serious defect.)

E4 The *reintegrated disintegrating* processes are Markoffian and appear in an environment of non-equilibrium.
("for a world that is always in equilibrium there is no difference between past and present, there is no history..." Robinson, *loc. cit.*)

---

[35] See Marglin (1984) for a lengthier discussion of the categories.

The introduction of axioms places some order in a discipline; it does not resolve issues or lead to new ideas. Bearing these limitation in mind the exercise has, at least, some paedagogical value. Unlike most economists, Koopmans (1957) discusses the problem of axioms extensively and astutely. He quotes Robbins (1935) who believes that economic postulates relate "to the way in which scarcity of goods...actually shows itself in the world of reality" or, that "the main postulate of the theory of dynamics is the fact that we are not certain regarding future scarcities."[36] The axioms here follow a similar, although more general approach: the uncertainty about future scarcities is one aspect of the coin; the other being that the "dynamics" of reproduction are themselves the originators of the ensemble of variations which encounter and respond to the field of scarcity according to their characteristics and capabilities. This outlook has a bearing in the following chapter where the Darwinian evolution is discussed in the context of the virtualities that potential variations of species offer for selection. It is not a random game but rather one where reproduction capabilities produce an ensemble of neighbouring variants on which selection operates. As to the axiom of scarcity, it may have been disputed by marxist (but not necessarily Marx himself) positions on the fullfillment of the species after the "integument [of capitalism] is burst asunder" (Capital: 929).

With the introduction of the stochastic element of capital, I avoid the "essentialist" argument referred to earlier. The above axioms can also be thought of as formulations for the

---

[36] See Koopmans (1975: 135) and his discussion of L. Robbins' "classical" work, *An Essay on the Nature and Significance of Economic Science*, Macmillan, London, 1935 (2nd ed.), pp. 78-80 of the reprinting of 1946.

being; Sartre is not explicit about the non-existence of alternative humanities in a non-scarcity, pre-hesiodian environment but argues that the one we are part of is the product of scarcity. Economists will surely subscribe to this axiom with few reservations. Marx, as well as Schumpeter, would have accepted the view that human progress is nothing but a flight away from scarcity through invention and innovation. In fact, the coefficient $A_S(t)$ of the production function in E3 above is the celebrated contribution of R. Solow (1957) which he has correlated with econometric data, showing that technological change was as important as capital accumulation in economic growth.[37] While an *ad hoc* coefficient which cannot be derived from first principles, it does have a strong heuristic value and it may be thought as the first attempt to introduce a dialectical approach to positive, mathematical economics. Sartre is taking the point a bit further, however, by emphasizing the human condition in the field of scarcity: "le champs de tension, de rareté." Thus, tension and *praxis* are tightly bound together in a dialectical order. The utilitarian neo-classical school expresses similar ideas through the utility of consumption $U(C)$ of a homogeneous sea of consumers, thus masking the "contre-homme" under the observable of unemployment and its statistical manifestations. By contrast,

[37] The mathematics should not hide the fact that in economics equations are the essential crutches for admissible argument which transcends verification and experiment. The representations which are assigned to various variables and functions are imprecise and subject to "rhetorical" interpretations. Solow's A(t) is a prime example of how imagination can be made convincing by supplying an experimental basis (data of economic activity) to a theoretical grid. Solow has been challenged on this but he has managed to subdue the advsersarial if only because his rhetorical ability was better. On the theme of rhetoric in economics see McCloskey (1983). One may consult G. Dillon, *Contending Rhetorics: Writing in Academic Disciplines,* (Bloomington, Indiana University Press 1990)and, applied to the sciences: "The Rhetorical Maesltrom," by Jan Golinski, an Essay Review, *ISIS* 1993, 84: 746-749).

the neomarxist and neokeynesian schools identify the struggling and competing social groups in the midst of scarcity. Of course, combining the implications of the dialectical with the "alterity" deriving from the inherent scarcity, and allowing for scarcity itself to become the need-developing mechanism (whether axiomatically or heuristically considered) makes for intractable ethics. As Aron (1975: xxi) writes: "indeed, man is defined by Sartre as a flight *out of the present into the future.* For the very activity that is knowing, we are reminded, changes that which would be known. This is, perhaps, the essential insight of existentialism: *That man retains the capacity to 'invent his own law'*" (my italics). This possibility conjures orgiastic revolutionary potentialities to scare the most enlightened liberal. One often wonders whether Sartre had been influenced by Schumpeter's work, transferring the ideas of the latter on economics to philosophy and a more basic biological behaviour: was it not Schumpeter (1934) who discussed the Phœnix-like process of *creative destruction* (= "reintegrated disintegration" of Sartre?) although confining himself to the mechanism of the market in directing the resources to optimal uses? Or, Schumpeter again, who suggested that capitalism is a "creed hostile to its very existence"?[38] Indeed, one may well conclude that a positive description of economics could be developed strictly on the dialectical lines if stochastic processes which are interweaved with the

[38] Aron in his *History of the Dialectic of Violence,* has been one of the severest critics of Sartre with an admitted *parti pris.* He set himself the task of studying the material exhaustively (line-by-line, as it were) only to impute to Sartre what the title of his own book implies: the dialectics of violence. But one wonders whether Schumpeter's description of crises imply less violence. Philosophy can be politics. But in Sartre we have found described many human mechanics of intertemporal economics — items typically avoided in econometrics and macroeconomics. I have profitted enormously from discussions on Sartre's CRD with Eleni Odoni.

nature of scarcity are taken into consideration. Note, however, that the word *scarcity* is nowhere to be encountered in the work of Rawls (1973) which is based on intergenerational principles of justice in terms of the formal neoclassical equilibrium terms; the perspective is strictly Kantian. Sartre's message, on the other hand, is suggested by the title of his work meant to contrast sharply with that of I. Kant's *Critique of Pure Reason.* While Sartre gropes with his theme with what appears to be a "sea of words," as some have implied, he attempts to describe a fundamental economic-dialectical mechanism and, in groping, he articulates the metaphor for the iterative procedure of the many-body problem in physics which leads to the description of the collective and of the independent behaviour of interacting particles. In CRD one recognizes allusions to the many-body theory as applied to economics. Kant, on the other hand, describes a being as if though the collective has no bearing on the beings themselves. One conjectures that in biological selection — right from the first molecular assembly — as well as in the economic world, the interaction of the collective with the "particles" is *the* strong, if not the only, mechanism for selection.

All the above are merely suggestive. While leading to no new algorithms, they inspired two new terms: the *Smithion* and the *Keynesion;* the *Smithion* (much like the physical electron, say) refers to the elementary, "undressed" economic actor in his total purity, while the *Keynesion* refers to the collective "dressed" economic actor much like the physical plasmon (the quantized plasma wave) or other many-body manifestations in physics. To observe the signals from the *Keynesion* to the *Smithion* we must then "look at the price system as such which is a mechanism *for* communicating information if we want to understand the real function ... The most significant fact about this system is the economy of knowledge with which it operates, or how little the individual participants need to know in order to be able to take the right action ... only the most essential information is passed on, only to those concerned" (Hayek, 1945). This is one example. Yet another is that of the Hegelian *contradiction*: in fact the concept of the *Keynesion* could not exist without that of the *Smithion*. Each acquires meaning only because of the other.[39]

Publicly, economists avoid the problems discussed here because they are unlikely to lead to algorithms with retrodictive (explanatory) or, better yet, predictive power. There is no professional advancement or recognition in such endeavours which are better left to philosophers. But there are insinuations that, in private, they recognize the heuristic value of these excursions in thought.

---

[39] The idea of the collective vs the individual behaviour is not all that new. Richard Whatley, the Archbishop of Dublin, wrote that "the transactions of each man, as far as he is let free, are regulated by the very person who is to be a gainer or loser by each, — the individual himself; who, though his vigilance is sharpened by interest, and his judgment by exercise of his own deportment, may chance to be a man of confined education, possessed of no general principles, and not pretending to be versed in philosophical theories; whereas the affairs of a State are regulated by a Congress, Chamber of Deputies, etc. consisting perhaps of men of extensive reading and speculative minds" (1894: 63, as quoted by D.N. McCloskey in "The Rhetoric of Economics," *J. Econ. Lit.* XXI 1983: 507). Recall too that Rousseau was differentiating the total community of the "general will" from the independence of the small autarkic family unit.

4. Summary

"Capital", as we have come to recognize, has various connotations in the mind of economists, historians, or the common economic actor. To reach a conclusion as to its origin, I have been led to plough a wide range of ideas in order to isolate two key functions of capital: the one refers to *virtual* capital and its *reification* while the other to its *dialectic.* The latter describes the process of reification in an environment of scarcity and non-equilibrium.

While studying these issues, I have posed some new questions although only suggesting avenues for their resolution. Specifically, I have proposed the study of the unexplored field of the *many-body economic theory.* Here I propose the connection with the corresponding physical theory where the *collective* and the *independent* particle (in economics I call it the "Smithion,") behaviours are quite distinct to an observer and depend on the latter's measurement and stick.[40] I have also introduced the concept of the economic quasiparticle, the "Keynesion," or quasicapital. The collec-

[40] Perhaps, the Hegelian and Marxist "contradictions" (Hegel, *Phenomenology of the Spirit*, p.399) implies what in physics is seen as the ambiguity between the collective and independent behaviour of particles. Depending on the measuring stick, a phenomenon can be observed in its collective behaviour (a wave, say) and in its independent (particle) behaviour. Expressed differently, "political economy thus demands that we move from the particular commodity to the market: from the viewpoint of the isolated producer to the examination of production and productivity as a whole ... the whole exists concretely and preexists in its elements; in one sense these elements are real, 'in themselves,' as moments of the whole, but in another sense they are simply abstractions in relation to the whole. The social whole is given as a practical organization of Praxis" (Lefebvre, 1958: 127). See also T.C. Schelling, *Micromotives and Macrobehaviour,* 1978: 13-14. In 1845-1846 Marx observed that human nature was "no abstraction inherent in each single individual" but rather "the ensemble of social relations," (K. Marx, *Theses on Feurbach,* Selected Writings, D. McLellan, ed. Oxford 1977: 137).

tive effects in physics, as in economics, derive from the two, three-, .. n-th-particle correlations stemming from the interactions; accordingly in economics not only is the "self-utility" $U(C)$ of neo-classical economy important — after all, "hell, is other people" (Sartre) — but also the higher order interutility functions $U_{ij} = U(|C_i - C_j|)$.

The economic process would hardly be interesting had it not been for the continued non-equilibrium, necessary to get the system in operation and mutating. In economics, the description of non-equilibrium is quite primitive. Much remains to be done to bring the concepts of Schumpeter, Robinson, Hayek, and the dialectics of Marx as interpreted by Lefebvre and Sartre, into contact and compatibility with the methodologies of the many-body theories and the mechanics of capital accumulation.

A program addressing the above problems, in order to develop economic equations of motion from first principles will surely prove to be intractable; yet the heuristic and paedagogical value of introducing such concepts as *quasi-capital, non-equilibrium* as the essential mechanism in driving the economic system, *interutility,* and *collective vs independent behaviour,* should not be discounted. Hegel's notion of blurring between observer and observed, is quite essential in the understanding of the economic process. Their interaction, examined in some length by Sartre in his *CRD*, supplies emotion into the otherwise dull capital theory. In a dialectical world, where computation of the future is an impossibility — a fact recognized even by Aristotle in his concept of ἐπιείκεια (the satisficing of Simon as noted in ftn. 43 below) — and where scarcity and the risks associated with it remain, there cannot be a capital theory in the model of gravitational theory, say. On the other hand the questions posed on the origin of the equation of motion in economics cannot be answered without the dialectical approach; what the equation of motion describes is not for-closed (predestined) by nature nor is it guaranteed to be forever.

This chapter does not touch on issues of distributed shares and of motivations, thus avoiding sociological implications. But it asserts that capital in its physical realization (and instrumentation in the field of scarcity) is the invention which makes *reality* and *intellection* (Sartre's expression) possible. In this non-ideological context capital is the essential instrument in economics. The management of the [physical] capital and of the capital supporting operations in the roundaboutedness of production, however, injects the issues of decision, property, etc. which, in a sense, sublimate the entity of capital and in the process homogenize it to an abstraction.

Hopefully this material on economics as widened beyond the perimeter considered useful by professional economists may be instructive in tackling the subject of the next chapter — biology. In between I felt the need to clarify in my mind some philosophical concepts and terms that are used throughout this book.

5. A Note on Philosophical terms

Philosophical terminology is a perpetual source of confusion. To assist the reader I adapt here a summary prepared by Piaget (1967) in the form of a table (II.2). Economics, the most dialectical of disciplines, is presented in the antireductionist column; the structure is there, given. The origin is never discussed. As a result economics is taken to be a phenomenological discipline. If the origins are to be explored, one can refer to the long-term prehistoric initial conditions and to the historic which Marx and other classify as dialectical. But what is the meaning of *dialectical* and of *reductionism*?

Dialectical Reason. Many have been puzzled with Sartre's introduction of the adjective "dialectical" to characterize reason. Among them, Levi-Strauss (1966) and Aron (1975) have attempted to interpret what Sartre had in mind when he qualified reason as dialectical. Indeed, there is one and only one reason, analytical reason, which is the currency of thought. (I do not dwell here on the concept of thought). Axiomatically, and in the realm of analytic reason, we accept a set of unprovables (concepts) and we build a logical structure such as to make statements and propositions come out true (i.e. compatible with experiment and experience). Kant, in his pursuit of the *foundations* (= Kritik, Critique) of analytical reason looked into how synthetic *a priori* truths are at all possible. The designation *a priori* is important because the assumption (or approximation, if you will) is made that indeed there are such *a priori* invariant truths independent of the analyst-observer i.e. of all possible experiment. This is what is designated as *idealism*, an issue we will encounter in subsequent chapters and in connection with Planck's views about the foundations of Thermodynamics or in connection with the process of measurement. Perhaps this is what Kant's *Pure Reason* conveys: *in* materiality to conceive of reason capable of (if not peremptorily) existing outside it as its externality. Although I do not follow a reductionist program to determine the physical (material) origin of reason, I nevertheless suggest that *Dialectical Reason* is the analytical reason that exists within materiality;[41] in a materiality which it can, nevertheless influence and alter at aggregate levels, while the components remain invariant. How this can be done is the thesis of Sartre and the content of intertemporal economics as I interpret it. The concept of the *dialectic* was introduced by Marx in the context of the Hegelian *contradiction.*

Reductionism. In practice we are all reductionists because this is the methodology which permits the isolation or, decoupling, of elements from the totality. This isolation (or, approximate isolation) leads to our understanding and to the formulation of laws and patterns. In physics, for instance, the outstanding results in quantum field theory (QFT) and quantum electrodynamics (QED) have been reached by a convergent set of reductionist steps which appear to have reintegrated the elementary in the totality (a possible way of articulating what physicists call "renormalization techniques"). In the study of the fundamental physical world, within analytic reason, reductionism is the key methodology that keeps improving the description of nature while guided by and guiding experimentation.[42]

---

[41] A question crops up here: if reason exists *within materiality* can the shift from classical to quantum physics affect logic? Redhead (1986) analyzes and then rejects Putnam's thesis that in view of quantum mechanics a new kind of logic has emerged. Of course it is still analytic reason that is at work. A Critique of Quantum Reason or Critique (Putnam) of Cognitive Reason (Searle) cannot therefore be justified.

[42] From the recent critique of the renormalization theory by Cao and Schweber (1993) that touches on the frontiers of physical research it is, however, apparent that this optimistic view is strongly disputed.

SCHOOL	Antireductionism structures, no genesis)	*Reductionism*	*Constructivism (Structures and Genesis)*
OBJECT	*Platonism*	*Empiricism*	*Dia ectics of Nature*
SUBJECT	*Apriorism*	*Nomilalism, Conventionalism*	*Historical Relationism*
OBJECT⊗ SUBJECT	*Phenomenology*	*Identification*	*Dialectic*
REPRESEN-TATIVE AUTHOR OR AREA	*Plato, Kant, Rawls, Planck, neo-classical economics*	*Descartes, Saint-Simon, Carnot*	*Schumpeter, Marx, Sartre, Dyson,* Solow,** Beckett,† Wheeler†*

TABLE II.2. Philosophical Schools, from Piaget (1967: 1240)

*To what extent can dialectics manipulate and use nature (within the physical constraints)? Dyson (1971) has given some technical limits for a practically perpetual possibility of existence, while Engels (in his *Dialectics of Nature*) has touched with much less justification the theme of a cosmological recurrence; much of CRD is a refutation of Engels! Dyson's world is a *dialectical* one if there ever was one and it is a version of the "Black Cloud" fictional brain of Fred Hoyle.
** Because of his contribution to the issues of technological change (1957).
†See chapter V.

At this level, there is no need for dialectics. Dialectics is not something competing with reductionism because the introduction of an extra dimension in the selection process namely *value* which has only intertemporal significance. It permits choice among alternative realizations (of physical or, ethical worlds) which are permissible in nature. Choice, or even a sequential set of random events with Markov statistics, leads to rules, laws, regularities and effects which cannot be accounted for with reductionist methodologies.

The procedure to affect the laws, and the effects at this phenomenological level of high complexity, is the content of dialectical logic. Within analytical logic, it is the logic of changing realities through the process of selection of certain realizations from an extended set of virtualities. Given that virtualities are many and not necessarily denumerable, the concept of *equity* (επιείκεια = satisficing) introduced by Aristotle and elaborated by Herbert Simon[43] preserves sanity and permits decisions to be made in finite time. Not all possibilities can be tried and we cannot know which of all the virtualities and their combinations will lead to "better" overall choices. Scarcity, whether of time allocated for decisions or the building of capital, is the guiding principle of dialectics as it is of intertemporal economics. Unlike the reductionist world, the dialectical world is not convergent.

Yet the reductionist method has its limitations. While the theoretical and experimental manipulative techniques have improved considerably and while the symbolics of physical representation have become accurate and consistent as a consequence of the mathematization of science, we are not one iota ahead of, say, Democritus in knowing the essence

---

[43] In his Rhetoric A.13.1374a.26, Aristotle was arguing that the law giver is incapable of foreseeing all the alternative situations and set a ruling for each. Had he attempted such a thing, "not even an aeon would have sufficed for enumeration [of the cases]." (*Καί ὅσα μὴ ῥάδιον διορίσαι δι' ἀπειρίαν, ... ὑπολείπει γὰρ ἄν ὁ αἰών διαριθμοῦντα.*) This difficulty then leads to the concept of epieikeia (fairness, equity but not in the sense of Rawls (1973)). That is, the freedom of the judges to interpret situations which are not exactly anticipated by the law or for which lengthy procedures could otherwise be required. We note that Hobbes in his translation of Aristotle's Rhetoric renders the passage as follows: "for an age would fail one in the enumeration," while he translates the word epieikeia as equity (example: "and equity is that idea of justice which contravenes the written law"). Could epieikeia be what Herbert Simon described as *satisficing* 2300 years later? For a review of some aspects of decision making in connection with Simon's work see J.G. March, "Bounded rationality, ambiguity, and the engineering of choice," *Bell J. of Econ.*, Autumn 1978. See also M.S. Macrakis, "Επιείκεια and Satisficing," *Ancient Philosophy,* **5**,1985: 53.

of it all. Mathematical expressions and indicators on instruments are still our phenomenological limits, and geometrical abstraction reigns supreme.

Reductionism has a normative significance in physics in that it demands *unification* i.e. the subsumption of all, in an all embracing construct — still a mirage. Remarkably, this heuristic has not failed to produce results in the world of the invisibles and of the abstractions. On the other hand, the cleavage between reductionism and dialectics precludes the thoughtless use of reductionist arguments for the formulation of policy, interpretations in biology (for which see next chapter), etc. Figure II-3 formalizes these divisions.

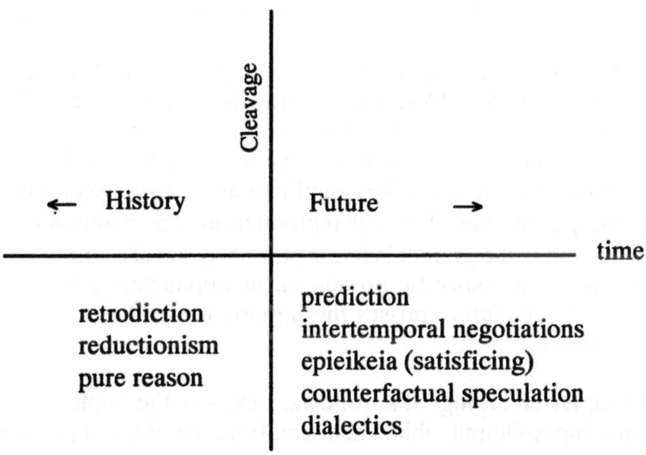

Figure II-3. A historian looks reductionalistically at the past events; he can only speculate about virtualities and outcomes then available; then he follows a dialectical critique.

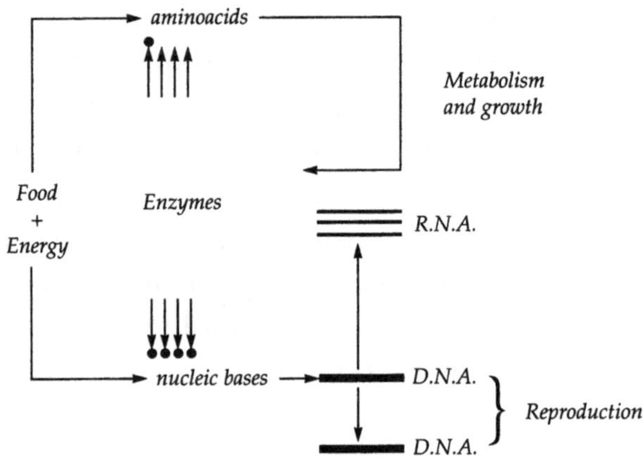

Figure III-1. Lwoff's schematic for the biological flow of energy.

This is Fig.11 and its caption from Lwoff (1962: 36) which describes the *flow* of energy for "Metabolism, Growth and Reproduction": "Food is metabolized. The small building blocks, the essential metabolites (amino acids and nucleic bases), are synthesized. Nucleic bases are organized into nucleid acids. The duplication of the desoxyribonucleic acid (DNA) is the very basis of reproduction. The manufacture of ribonucleic acid is the prerequisite for protein synthesis. amino acids will be patternized into specific protein on th ribonucleic acid template." This type of accounting misses the importance of the structure in the form of capital.

In the chapter on biology that follows, I explore the implications of an economic Input-Output table when seen from a biological perspective.

CHAPTER III

BIOLOGY

ἄν πολλὰ βάλλῃς, ἄλλοτ' ἀλλοῖον βαλλεῖς[1]

1. Introduction. Biology 101
(The Economics of Biology)

RECALL KEYNES's aphorism, to the effect that

> it is by reason of the existence of *durable equipment* that the economic future is linked to the present.

This can be extended to apply to all of biological systems. The maintenance of biological order is in fact conditional on the existence of "durable equipment." In some approximation this principle is also implied by Watson's assertion that "*Life* itself emerged when primitive nucleic acid and protein became spontaneously *encapsulated* in lipid-containing membranous sacs that provided an environment where organized growth could occur."[2]

---

[1] "If you throw [dice] often enough, your luck will change." Aristotle, 463b21.
[2] Watson [1977: 456]. My italics.

I have used the quotation simply as a point of view. It is difficult to accept it completely, however, as it already implies a built-in synoptic model for the origin of life: that the formation of the cell through the *spontaneous encapsulation* of preexisting primitive genes was the only outcome possible. It may be more prudent to pose a question as follows: "which came first, a functioning metabolism protected by some kind of membrane against dilution and destruction of surroundings, or a large molecule that survives by making copies of itself from materials in its surroundings?"[3] In other words, which is older? The *protobiont* or, the *naked gene.* In summary, two schools exist on the fundamental question relating to the steps from elements to structures. One, supports the view that life originated in the gene which subsequently appended around itself auxiliary structures. The other considers that the gene itself was the product of a long evolution of primitive organisms with the implication that its components may have had other biological roles in the primitive structures before becoming the key to replication and heredity. In simpler terms the problem is often captured by referring to the chicken and egg paradox.

If in his celebrated speculative essay *What is Life?* Schrödinger[4] logically inferred the necessity for a code-carrying chemical substance—later identified as the double helix DNA macromolecule—he nevertheless left the mystery unresolved

---

[3] R.E. Dickerson, "Chemical Evolution and the Origin of Life," *Sc. Am.*, 293, 3, p.76, 1978.

[4] E. Schrödinger, *What is Life? The Physical Aspects of the Living Cell,* Cambridge, 1945. I note here that the impact of this publication on the development of molecular biology is often disputed by various professionals and historians of science. To get a feel consult: E.J. Yoxen, "Where does Schroedinger's "What is Life?" belong in the history of Molecular Biology?," *Hist. Sci.,* **17**, 1979: 17-52 and Keller, Evelyn Fox, "Physics and the Emergence of Molecular Biology: A History of Cognitive and Political Synergy," *J. Hist. Biol.*, 23, 3, 1990: 389-40.

while only suggesting that, out of chaos, complex molecules will assemble. Indeed, half a century later we still are far from any suggestions leading to some philosophical concept, a physical theory, or a technique explaining what life is and how it came about. In between a great breakthrough occurred: the discovery of the DNA double helix by Crick and Watson will surely remain in the annals of science as a unique and fundamental contribution providing biology with a unity, perhaps, even transcending the Darwinian synthesis itself. Yet, the helix takes off with a given structure, theory, and practice which provide no clues as to its origins or its role in the observed cellular organization. It is not known whether a cellular membrane is self-priming (autocatalytic) or completely coded in the genes. There is no record left of the early stages of the cell formation since it has been erased by the ones following. I avoid the adoption of any of the various propositions and tentative schemes which one encounters in the literature; I limit the discussion to a model which may prove practical and, hopefully, fertile. Of course, by studying the ideas of Eigen[5] and Küppers,[6] one might be led into believing that to start a biological (reproducing) assembly, one only needs a "soup" containing all the ingredients. However, biological experimentation and engineering still depend on the use of natural structural capital (the cells) or its products, and the effort is directed towards the identification and elucidation of various processes rather than the overall mechanism for the origin of reproduction.

---

[5] M. Eigen, "Self-organization of Matter and the Evolution of Biological Macromolecules," *Die Naturwissenshaften*, **58**, 1971: 465-532.

[6] B. Küppers, "Towards an Experimental Analysis of Molecular Self-Organization," *Die Naturwissenschaften*, **66**, 1979: 228-243. B. Küppers, *Molecular Theory of Evolution*, Springer and MIT Press, 1990.

These concerns led me to the development of a biological model along the lines of an economic physical capital (engines, say) in interaction with energy. Being an accounting model it does not resolve the issue of the sequence of events that led to the phenomenon of life. Nevertheless, it does impose constraints on what can be realized (and not just on conservation of matter and energy). Unfortunately when one tries to fill in the detail, the job becomes too difficult. Yet the exercise is not to fill in the I-O table as if it were an economic transactional world but, rather, to illustrate and stress the mechanics of roundaboutedness. These deeper issues are, however, beyond the scope of my essay. Rather my objective is to suggest that whatever mechanism is finally established it will have to fit the structural constraints pertaining to capital "reproduction." The accounting of energy flows in biological systems has been discussed by many authors before; see, for instance, the work of Lwoff.[7] But this earlier method of accounting misses the importance of the three-dimensional structure in the form of "physical capital" as well as the process of roundaboutedness.

To keep track of these mechanisms I now sketch a possible mapping of the elementary biological process of the cell by adapting the Input-Output (I-O) table which was discussed in the previous chapter and to which the reader is now referred. The I-O table of the representative cell or, multicellular organism, may one day be fully described although the amount of detail and information necessary will be forbiddingly large.[8] Yet its simplest, basic form, has the structure shown

---

[7] A. Lwoff, *Biological Order*, MIT Press, 1962: 36, figure 11, "Metabolism, Growth and Reproduction," describes the flow of energy and has a caption reproduced here on p. 60. This type of accounting, however, misses the importance of the structure in the form of capital.

[8] While the development of I-O biological tables may prove forbidding (in fact even in the macroeconomic field considerable detail is missing and

in Figure III-2.[9] The 2 × 2 matrix would then correspond very grossly to the activity, of the biological process where energy and materials are the principal inputs to replication and sustenance.

	Mat.	*Met.*	*Cons.*	*F.D.*	*Exp.*
Materials	a_{11}	a_{12}	C_1	I_1	
Metabolism	a_{21}	a_{22}	C_2	I_2	
Capital	K_1	K_2			
Labor	L_1	L_2			
Imports					

Note: $\sum I_i = I = \sum \dot{K}$

Figure III-2. An Elementary Input-Output table with 2 basic biological sectors, Final Demand (C,I), Capital, and Labor requirements along with "imports" and "exports".

Building on the idea of the biological capital as constituting an engine of reproduction, we may be justified in developping the I-O table even further. Could the sectors be characterized with more specificity and in more detail? And if so, what should the rules of characterization be? Can we, per-

---

what becomes available is always too late to be of any comfort to planners), it is possible to expect that studies of evolution of primitive reproducing structures with an I-O methodology will prove valuable. A table of say 100 inputs studied before and after a mutational change can be a good way of measuring and establishing substitution possibilities and changes in coefficients, energy requirements, etc

[9] Adopted from E.A. Hudson, and D.W. Jorgenson, "U.S. Energy Policy and Economic Growth, 1975-2000," *Bell Journal of Economics and Management Science,* June 1974.

haps, look into some important, well described processes for suggestions? For instance, according to the *Central Dogma*[10] of molecular biology as formulated by Crick, genetic information can be transferred autocatalytically from DNA to DNA or, catalytically from DNA to protein, but not from protein to either DNA or RNA, at least at the current stage of evolution.[11] Because of the central dogma, some of the coefficients a_{ij} will be zero. For instance, $a_{PR,DNA} = 0$ and $a_{RNA,DNA} = 0$ is used in the enlarged version of the I-O table of figure III-2. Tentatively then, I map the "intermediary" process in the biological system and, to submit to the *dogma*, I assign a key position to DNA. DNA is the information sector which provides the services and the guidelines for the manufacturing sectors. Simply, "the structure of DNA gave the concept of the gene a physical meaning by which all properties can be interpreted. Most important, DNA — right there in the physical facts of its structure — is both autocatalytic and heterocatalytic. That is, the genes have a dual function, to dictate construction of more DNA, identical to themselves, and to dictate construction of proteins very different from themselves."[12] Most authors stress the centrality of DNA in the biological models. However, the implications and the principal point of this note are that DNA is real-

---

[10] The *Central Dogma* in molecular biology, of course, is simply a restatement of Weismann's doctrine, applied explicitely to intracelular molecular dynamics. Weismann held that the *soma* is formed from the transmission of the hereditary elements of the germ line. But "if the Weismann-Central Dogma description of biological systems (all organisms, from prokaryotes on up) is as accurate as it seems, there is an asymmetrical vector of causality within organisms, one that reinforces the common belief that DNA-RNA are in some sense more 'important' than economic matters in biological systems. Genetically based information is primary; it controls the elaboration of the economic systems on which it depends, ironically, for its own existence." (Elredge 1989)

[11] For an RNA-directed synthesis for instance see H.M. Temin, "RNA-Directed DNA Synthesis," *Sc. Am.*, **287**, Jan. 1972.

[12] H. Judson, *The Eighth Day of Creation; Makers of the Revolution in Biology,* Simon and Schuster, 1979: 23.

ly only a cog in a very involved process of roundaboutedness.

	PROTEINS	*DNA*
PROTEINS	$\rightarrow \bullet$	$a_{PR,DNA} = 0$
DNA	$a_{DNA,PR} \neq 0$	$a_{DNA,DNA} \neq 0$

Figure III-3. The *Central Dogma* illustrated in the form of an I-O Table.

What could be the reasons for the specialization observed and the one-way interaction paths between DNA and RNA? The RNA molecule receives instructions from the DNA and then provides the templates for fabrication of proteins which in their turn control metabolism in a cell. The process suggests that the "master template" (the DNA) to survive as a molecule must keep its distance from molecules in the cellular protoplasm that are capable of breaking it up. Separation of the DNA from the metabolically active cytoplasm allows for long-term informational stability of the cell as a whole. The replicational repeatability is thus attributed to the structures that developed which precluded strong chemical interactions between code-carrying molecules and molecules with metabolic qualities.

As an illustration of the intermediate processes I sketch a more "realistic" I-O table in figure III-3 in which the reader can allow his imagination to roam freely. With its support, one may discuss qualitatively the input of energy which in a biological system undergoes many transformations and is transported long distances before it becomes available for the *final demand* and the development of *work*. I look at the cell as a thermodynamic engine whose primary function is to develop work.[13] Yet, on specializing, cells do contribute to the large organisms many intermediate functions of "planning," of bringing resources (however minute in quantity) to specific places and times, in coordinating, etc. They contribute the overhead management functions required to make complex systems work coherently. As I discuss in Chapter IV, *work* — whether required to add a molecule to a molecular system or, pertaining to the function of cilia or muscles — is the measure of value in energy transactions and the key to the understanding of the second law of thermodynamics. I must, then, introduce an energy transformation and processing sector which would have to include the production of energy-rich ATP molecules, and also an energy transportation sector. For completeness, I must then include a sector for the transportation of energy with the use of metallobiomolecules (i.e. haemoglobin, haemocyanin) which are binding molecular oxygen to carry it across long paths towards its end utilization. A truly formidable exercise. However, the following remarks could be of some help for the justification of the selected I-O sectors. Note that the primary energy inputs into biological systems are photons and materials (carbohydrates and fats). Through photosynthetic or oxydative phosphorilation, these energy inputs are transformed and stored in the form of adenosine triphosphate (ATP), the universal "accu-

[13] The elemental capital in biology is the cell (in interaction with neighbouring cells). It differs in structure (not function) from a thermodynamic engine. The latter is a synoptic device with reduced number of degrees of freedom (pistons, boxes) liable to "all or nothing" operating capabilities (about which more in Chapter V). But the assembly of these engines is done in the intermediate sectors of the economy which utilize energy, materials and then supply the machine as an investment (I) which then becomes part of the production system.

mulator" of energy in the bisphere. The ATP molecules are stored in cells so that they may be triggered (switched) on demand to promote various functions such as muscular contraction in pluricellular systems, the synthesis of molecules, etc. Specifically, it is thought that the rod-shaped mitochondria in cells are sites of cellular respiration while in plants it is the rod-shaped chloroplasts which convert the energy of light into chemical energy of ATP. The hydrolytic dissociation of ATP in water solutions releases energy which results into the formation of adenosine diphosphate (ADP), the freeing of P_i ($ADP \Leftrightarrow ATP + P_i + .4$ev), the inorganic phosphate, and, the preparation for the recycling of ADP back to ATP. From this simple example one can view the complexity that can rapidly overwhelm the effort to synthesize a biological I-O table.

I note further that in a practicoinert (static) economy,[14] a standard I-O representation would suffice since all coefficients would be time-invariant. However, to provide elastic coefficients which could accommodate evolutionary change, we must now look into trade-offs between consumption and "investment" as well as into the assignments of capital functions among the various types of capital; and, at the level of a cell we cannot separate what is physical capital (the processing machines) from what is referred to as labor in macroeconomics. Additionally, to provide for flexibility at the "micro"biological level, I introduce an error inducing machine

14 The "robinsonnade" according to Sartre (1960: 642).

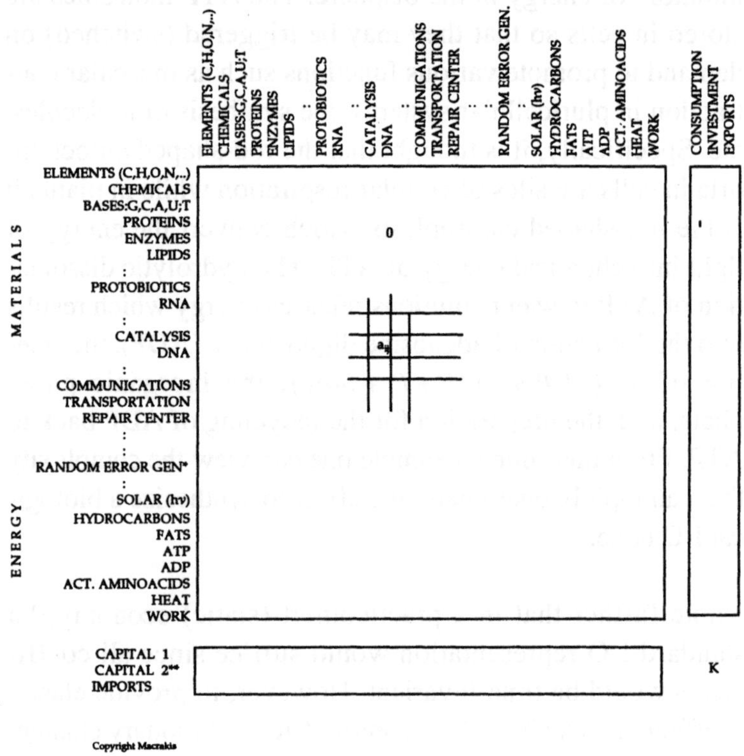

Figure III-4. Towards the development of a biological I-O Table. The description implies a generalized biological system which includes as subsets elements that cover the range from a simple bacterium to a human body.

*The random error generator could be included in the sector of imports if outside the biological system.

**Capital 2 represents labor in macroeconomics.

with a random number generator; while the sequence of events that produced the cell remains obscure, stochastic processes[15] add a component to the I-O table which only the price mechanism and the Smithian invisible hand can provide in the description of the macroeconomic processes. Referring now to the DNA-RNA transactions discussed earlier, I sketch in the I-O table of figure III-3 some vertical lines to suggest that, as the process of replication is influenced by errors, there result mutant outcomes.[16] The distribution of these is not specified or known, but Eigen[17] speculates that it is discrete and that it results from and is responsible for the process (and principle) of natural selection.

---

[15] See R.C. Merton, "An Asymptotic Theory of Growth Under Uncertainty," *Rev. of Econ. St.*, 1974: 375. But also consult J. Steindl, *Random Processes and the Growth of Firms - A Study of the Pareto Law*, Hafner, 1965.

[16] In addition to exogenous effects one can also include "feed-back" responses which may lead to reprogramming of the DNA sequences in the cells of the immune system.

[17] "Self-replication and Molecular Evolution - Selection and Evolution Governed by Natural Law," paper presented at the centennial of Darwin's death in Cambridge, England, 1982; I thank Prof. Eigen for sending me a preprint of his lecture.

2. *Biology 201 (The Dialectics of Biology)*

In the previous chapter, I discussed the phenomenology of capital and its accumulation posing a number of questions concerning its physical origin; here I suggest that the origin of capital must be sought in biology and that the I-O table could serve as a heuristic conceptual frame for the characterization of (an invariant) structure of the biological world. This naturally brings one to the questions of the origin of life and of its primitive (monocellular) beginnings, a popular subject which has acquired prominence with the support of modern advances in molecular biology. The resulting literature is by now vast, often technically difficult, and as might be expected highly speculative; it has also characterized the various schools of thought. More recently there have been attempts for unification in the field by providing a new synthesis which extends the neodarwinian evolutionary theory (i.e. the coupling of Darwin and Mendel) of the Dobjansky school to the molecular level; this feeds back from the molecular to the macroscopic.[18] To order our thoughts, however, I indulge in a short review of the various schools which are differentiated by their models for the origins of "life." Two are the basic models which have been proposed. One assumes the pre-existence of replicative chemistry and that naked genes were spontaneously encapsulated to form

[18] A model presentation in the sense of combining fundamental microbiochemical considerations with a sensitivity to the phenomenology of biology is that of Eigen (1971, 1981a&b). The conclusions are still speculative and often hard to adopt, yet, I refer the technically oriented to this uniquely coherent and modern presentation which is also accompanied by an extended bibliography.

cells (Watson, 1968) with the implication that the genes acquired a throwaway machine for their own utilitarian purposes — utility of replication U(R)! or, in simpler terms: organisms (the *phenotype*) are DNA's way of producing more of DNA for the "preservation" of the aminoacid sequence. This is the *genotype* model for the origins of life (i.e., the modern version of Samuel Butler's aphorism — the chicken is the egg's way of making another egg). The other model assumes the preexistence of large assemblies of proteins in the form of coacervates which developed a replicative technique later formalized by DNA as the vehicle for information storage and transfer. In simple terms, the implication of this model is that the egg is the chicken's way of making another chicken — the popular view. This model is referred to as the *phenotype* theory of the origin of life. Its axiomatics require that the origins of life preceded the origins of heredity; it has important implications on the mechanism for transfer of biological information from proteins to proteins and then from proteins to genes — in variance with what is presently the observed sequence.[19] The *phenotype* is the phenomenological human being, i.e. a machine in interaction with the environment and other machines.

Note that selection is not tested at the replication level (where information is stored) but at the phenotype performance level i.e. the large scale cellular or multicellular organism where the "evaluation" of the genetic messages is performed. Thus the link between genotype and phenotype or, alternatively,

---

[19] The phenotype model implies molecular Lamarkism which will cover a broad spectrum of theories (from those of Oparin to Lysenko). For a critical review see Barbieri (1981) who suggests a third model based on the ribotype. He claims that the economic accounts of cells indicate that most of the energy and materials are invested in the ribosomes rather than in the information storage devices suggesting that ribonucleoproteins created the cells and that the cell is just the instrument for their replication.

between information and its expression is intertemporal and hidden in the complexity of the roundaboutedness described in the I-O table. It is the characteristics of the phenotype that permit an organism to survive in some given environment; the phenotype is considerably more elastic in resource utilization than the rigid genotype.[20]

Genes occupy a three dimensional space yet it is the sequence of their chemical components (A,C,U,G) that constitute the memory store of the information and the instruction manual for translation. By contrast, the 3-dimensionality of proteins is essential for the structural and other biological functions described in the I-O table. The table, as such, will not describe the modes of interaction between the information (genotype) and the structural (phenotype) space. Thus, even though the *Central Dogma* asserts that information flows from genes to proteins — never the other way — it does not intimate why and how the flow of information is transmitted, or how the process is initiated. One can only speculate, however. Monod (1972), for instance, proposed a precellular evolution occurring in two phases. A first phase during which time an inventory of nucleotides and aminoacids was developed, and a second phase in which macromolecules capable of replication appeared on the scene (no mention of "naked" genes). These phases are then followed by the critical third phase where primitive replicating molecules encapsulated themselves (but then who built the capsule? the genes, perhaps?). Monod is quite explicit when he states that "replicative structures" preceded the "teleonomic apparatus" that is the cellular complex and, thus, places himself squarely on the side of the *phenotype* school of ori-

[20] "the realm of life is at the same time rigidly unified abstractly and immensely diversified phenomenologically" (Chargaff as quoted by Judson, 1979: 222).

gins. But then, in the same breadth, he contadicts himself by referring to the "permeable membrane without which there can be no viable cell."[21] He, nevertheless, admits that his hypothesis of the "gradual [non-spontaneous?] emergence of the teleonomic systems around replicative structures," brings one to a "real *sound wall*" because no one has any "idea what the structure of the primitive cell might have been." A hypothesis is thus dodged. Eigen (1971, 1981) in a series of noteworthy contributions, brought the *genotype* model some steps away from a hand-waving argument by proposing specific theoretical models, partly suggested by experimental findings. He too belongs to the *genotype* school and bases his argument on chemical reasoning and the abundance of the chemical species. His models are vague as to the origin of the cell itself, however. "The innovention of DNA," he writes, "made possible the formation of the individualized cells in which the cell is synchronized with the replication of its genetic material."[22] Later, he and his collaborators (1981) speculate and suggest that transfer-RNA might be an early gene because it fulfills certain preconditions. Eigen's work on *noisy* self-reproduction of molecules implies, however, a biology without cell boxes as the precursors to a cellular biology.

The genotype model surely begs the question as to the development of the membrane for the "spontaneous" encapsulation without the benefit of a stored program. This issue is not cleared up in the Eigen papers who asks why life is everywhere cellular and he sites the advantages of the cellular

---

21 Monod, 1972: 140 & 143; Eigen, 1971; 516-517.

22 Incremental steps do not necessarily imply that the evolutionary process is continuous in time. There may be long periods of stasis (the practico-inert) followed by a rapid shift to another evolutionary state, the domination of a particular evolute.

organization as the "protection from fluctuations in the external environment and maintenance of concentration gradients ... cellular organization was needed because it was the only way to solve the one problem of information processing in evolution that self-replication and cyclic cooperation were unable to address: the evaluation of information in genetic messages."[23] Apparently here Eigen follows the neo-darwinian synthesis which, as we have seen, requires the phenotype for the "testing" of the impications of any innovations in the genes. Yet one may object to the language and formulation used because it implies that there is no alternative from reaching what is observed or, that there is a manipulator with a utility function (in the sense of Bentham) who makes sure to evaluate and optimize. These deontological requirements (self-optimization by adaptation, for instance), at the early phase of replicational history, are questionable. What is needed is further research and experiment as well as a cleansing and sharpening of the vocabulary which, being so anthropomorphic, often misleads. Indeed, the language used in the biological literature addressing the origins of life at the molecular level often reflect utilitarian, hedonic and selection values which are observed in the phenotype (i.e. the phenomenological being); the implication is then that every model at the level of the phenotype has a corresponding model at the molecular level. Endowing molecular structures with Darwinian principles provides an incomplete picture. We may believe that nature is not anticipatory, that there is a Smithian invisible hand, that the role of chance is overriding, yet it is the dialectics at the phenotype level which endow all these microscopic events with utility.

The dialectical principle of evolution is now known to extend down to the level of the macromolecular complexes — the

---

[23] Eigen, 1979: 516-517.

building blocks of organisms; the evidence for this resulted from the study of the aminoacid protein sequences from different species. One of the well-studied sequences, for instance, relates to the evolution of globins and its tight correlation with the evolution of species. Thus, any conjecture about origins must resolve the problem of replication within the context of evolution. Accordingly it is unlikely that any synoptic, spontaneous, event structured the first functioning cell. In fact, all studies in the evolution of biological macromolecules convince one of the potentiality and potency of this striking mechanism and of the need to incorporate it as a basic principle. Evolution implies deviation from perfection in repetition (stasis) and, therefore, the need for errors which provides virtual states. It also needs a testing-proving field (i.e. interaction with other macromolecular ensembles).[24]

A review of models for the origins would be incomplete without reference to alternative models in evolution. Woese (1981), for instance, suggests (the obvious) *progenote model*: that the cell itself was the result of an incremental evolution from a simple, crude, noisy, error-prone translation apparatus with "a small complement of genes...to a complex genetic apparatus" rather than a synoptic event suggested by the spontaneous encapsulation model. The extremely high accuracy of replication could not have evolved in an open environment of a pre-cellular phase because the outcome, i.e. the cell, is the being tested for its life cycle properties.

After this short discussion on the various schools on the *origins* I get on with the next task which refers to evolutionary guidelines. The aphorism "selection of the fittest" is tautolo-

---

24 A.C.T. North and J.E. Lydon, "The evolution of biological macromolecules," *Contemp. Phys.* **25**, 4, 1984: 381-393.

gical (Stent, 1969); it will not do. But then what, if any, is the guiding principle in evolution? "Some say: ... [the] economization of free energy. [But this is] not true universally ...The use of information associated with high selective value rather than economization with the consumption of free energy is the decisive factor in evolution."[25] What is needed, Eigen adds, "is a general theory which includes the origin of self-organization of ('valuable') information." His argument in support of this position is "that the process of selection usually includes an economization with respect to the information content. In a fixed environment those sequences requiring the lowest number of digits in order to include the necessary information will usually show the highest selective value." In effect Eigen proposes what might be labeled the *information theory of value* in the tradition of Marx and Lotka: Marx had proposed the *labour theory of value* and Lotka what amounts to the *energy theory of value*. Ahead of the "energy crisis" of the seventies and the related literature, Lotka[26] had proposed energy as the decisive factor in evolu-

---

[25] Eigen 1971.

[26] Lotka (1924: 357) was concerned with the availability of energy: "what must be the effect, upon the world-as-a-whole, of the general scrimmage for available energy?...animals are essentially catalysers, oiling the machinery, as it were, and assisting energy in its downhill path to levels of lower availability (higher entropy), and so long as there is an abundant surplus of available energy running to 'waste' over the sides of the mill wheel, so to speak, so long will a marked advantage be gained by any species that may develop talents to utilize this 'lost portion of the stream.'" Lotka then continues with the observation that "such a species will, therefore, other things equal, tend to grow in extent (numbers) and this growth will further increase the flux of energy through the system. It is to be observed that in this argument the principle of the survival of the fittest yields us information beyond that attainable by the reasoning of thermodynamics." Here, we have once more the recognition of an important effect which was described by F. Braudel in *Capitalism and Material Life 1400-1800*, Harper Torchbooks, transl. M. Kochan, 1974. In the Chapter "L'épargne-bois" (p. 217) he writes that "having become more efficient, hearths increased astonishingly in number." In p. 270 he also writes that because of the increased demand, more trees were felled and that this led to the depletion of the forests! This remark is very significant for the current times characterized by conserva-

tion. Tentatively he enunciated the "law of evolution adumbrated as a law of maximum energy flux." One should be skeptical about the adoption of such a law since the utilitarian approach to consumption (maximizing discounted utility of consumption *U(C)* — discussed in Chapter II) includes the law of maximum energy in subtle ways. It is not the energy or, the entropy theory of value that is relevant in biological evolution — and, by extension, in macroeconomics. Efficiency in energy utilization takes a back seat to increasing growth rates which in "competition" allow winning out over sluggish users. Thus, economization of free energy, as such, cannot be the guiding principle in biology; it is not a guiding principle in micro- or, macroeconomics and there is no reason to believe that the principle is valid in archaeobiology. This view coincides with current trends in evolutionary biology that emphasize non-selective forces in evolution.

Recall now the discussion in Chapter II about the economic production function with the two variables — generalized capital and materials — needed for an economic output; value can only be shared between these two factors of production. A biological machine — however elementary — will have to submit to related rules. In its "self-design," a biological machine behaves *ex post* as if though there has been a competition between the cost of its own formation (capital) and the cost of its running (energy in the form of work or its equivalent). The production function takes the simplified form

$$Q = F(K;t) = \dot{K} + C; \quad (\dot{K} \equiv I)$$

---

tion and environmental concerns: "other things" stay equal in our computations but not in nature's ways.

where the output Q goes for consumption (energy →waste heat) and for investment, that is for building the next generation of cells (materials, translation, replication). The various possibilities to allocate the output beween investment and consumption could form a Consumption-Investment frontier

$$Q = (aC^b + (1-a)I^b)^{1/b}$$

known in economics as the *production possibility frontier.* The implication is that within some limits there is the possibility for various biological experiments which land on the C-I frontier. There may now be a temptation to introduce a discount factor and, correspondingly, a utility for reproduction. But this will cary us back to the information theory of value proposed by Eigen.[27] Note that acquisition and use of information on capital and its rate of change is an intertemporal ("planning") process; this process acquires value and meaning at the higher levels of organization where negotiation with the future becomes possible. Stated differently "our biological energies cannot be determined only by the past of the species, but also by the future they contain in them."[28] At the level of the cell, the physics are not (thought to be) anticipatory; being locked in an inelastic performance, with occasional possibilities for mutations, it is the *ex post* "assessment" which becomes the measure of repeatability of a replication experiment.

The discussion, this far, had as an objective to justify my insistance on the importance of *capital* at the cellular level and

---

[27] *loc. cit.* Information transactions (storage and retrieval) are costly and consume time and resources. But as noted by Barbieri (1981), the cell invests much more energy and material resources in its ribotype than in its genotype.

[28] H. Lefebvre, *Dialectical Materialism*, 1958: 114.

on the identification of the "cell-box" as the origin of the economic capital. Nevertheless, the mechanics of the encapsulation remain a technological puzzle. The sequence of chemical and catalytic steps which resulted in encapsulation may never be traced, unless shown that only a very limited number of steps can uniquely lead to the results. Speculation cannot cease. Nevertheless, tracing the incremental steps appears to be beyond the capabilities of science, as the alternatives are many, while archaeobiological evidence has been erased.

The formation of *capsules*, i.e. permeable membranes separating two regions of space one of which may even contain a primitive genetic coding molecule, can be visualized as having originated in various ways: from coacervates, from microspheres consisting of lipids or protenoids with hydrophobic side-chains which can spontaneously assemble under conditions of aminoacid condensation (Fox 1965, 1973), but certainly not in any coherent manner. These "cell-boxes" can, nevertheless, perform, erratically and unprogrammatically, a number of energy and materials transactions. And, if specific molecules such as ATP are encapsulated, then, even energy-driven volume variations can be expected. At this point specific mechanisms must be invoked. This brings us to observe a very interesting fact: all work in biology, and this includes the origins of cells, has been thought out in terms of "classical" structural chemistry and thermodynamics. Of course current chemical thinking rests on quantum mechanical foundations on bonding. But working chemists do not have recourse to the quantum except, perhaps, when dealing with rather elementary interaction systems. Despite this distancing from the quantum, the models and metaphors that ensue have been quite adequate in chemical work. It is only recently that quantum mechanics is invoked for thinking on thinking; it is now conjectured by some advocates[29] that the elusive *consciousness*, itself, is a product of quantum mechanical processes (yet to be specified in any detail). There is, then, a wide unexplored region between the quantum world of physics and that of the mind. Both sides are tackled; it is the intermediate situation that just fails us, as the more "mundane" problems of quantum processes in nerve, membrane, and muscle physics, of which there must an abundance, are still in infancy and conjectural stages. Physicists apparently can develop and handle deep unified theories on matter operationally successful up to high energies or theories on the solid state and the many-body problems, of astounding effectiveness; yet, the complexity of biological structures, that entails the comprehension of a system at work, evades them. Were we at this point to introduce the quantum, we might immediately run into a philosophical, conceptual and technical quagmire. Recall that in Bohr's view biology and quantum theory are at odds. If "experiment" is the key element in quantum mechanics, then an experiment in biology excludes "the possibility of maintaining the organism alive."[30] Suffice to say that our understanding of the structure of DNA, for instance, has been obtained through a lenghty sequence of events: the X-ray measurements, that provided the information for its helical structure, were done on a crystallized sample, containing a vast number of molecules separated from their cells through various mechanical and chemical processes; the test-sample, thus derived from an enrichment process, had no connection with a biological structure from which it originated. Most biophysical experiments are performed in such a manner.

[29] This will be discussed in Chapters V and VI.
[30] N. Bohr, *Atomic Physics and Human Knowledge,* Wiley, NY 1958: 20.

But if electrons produce interference effects, as they go through the proverbial "double slit" and their wavefunctions sequentially collapse as events progress, can one imagine what is going on as an electron or an atom traverse a membrane in nerve or intestinal walls? or when two membranes are close to each other? or when a macromolecule unfolds? Indeed at the level of the cell, quantum effects would be prevalent in the mechanisms of interaction of free floating macromolecules, molecules, atoms and electrons, with membranes. It is expected that the micromechanics of the operation of the muscle will, one day, be described with exact quantum mechanical techniques that describe the forces on a molecule boxed between 'rigid" walls. This refers to a model, discussed in Chapter V, which is closely connected with the most elementary exercise that students learning quantum mechanics encounter. One can then speculate that the same quantum mechanical wall movement can be invoked for the assembly of large, extended molecular membranes into enclosures resembling cells. Consider, then, two membranes roughly parallel to each other embedded in an assembly of various chemicals and ATP molecules. Triggering of the P molecules to release energy and to free a phosphorus atom imposes a force on the boundaries of the box as the wavefunction changes while the two membranes approach each other. This is a mechanism that can be invoked for the micromechanics of muscle contraction; it might lead to a "spontaneous encapsulation" of the ADP, ATP, P_i, fluids, and whatever materials may happen to be there at the time. In view of Murphy's law in chemistry to the effect that "anything that can form under appropriate conditions will do so,"[31] it is generaly accepted that complex molecules and proteins — simple and in the form of coacervates of mem-

---

[31] Eigen, 1971: 471.

branes — were synthesized before the origin of the cells; but not that the original (the first) *cell* could be assembled by systematically knitting itself without a program (deriving from the instruction manual of the genes, say). On the other hand, the assembly of a cell-box, while not resulting into a replicative system, does have the possibilities for the performance of some elementary functions: retaining a thermodynamic non-equilibrium, transport of chemicals through membranes, volume variation, internal synthesis of chemicals, interaction with photons and ionic media, contraction — a precursor to muscular *work*, etc. Indeed, one may conceive of the construction of a molecular non-replicating engine operating, and even stochastically evolving, in both content and performance (Woese, 1981).

Encapsulation (whether spontaneously or incrementally completed) allowed for the motility of the apparatus which could transfer itself by responding to various gradients (of acidity, heat, etc.), developing celia — flagellar motors — which increased the probability of sustenance by exposure to undepleted resources of energy and materials. The functional performance of elementary cells (thermal, materials and energy transport; microscopic motility, contractions — the muscular beginnings) is what encapsulation achieved. These functions may have changed the haphazard replication of molecular strings which result from a dash of frenzied activity to a "managed" coherent activity and, on the way, established the economic physical capital.

An End Note: Note that in the macroeconomic accounts, the I-O table plays the role of the genotype in which all the instructions for the transactions are stored. The phenotype, on the other hand, would correspond to the collective physical capital and labor. So that, even in the macroeconomic world, we encounter the chicken-egg paradox. Indeed, it is not clear whether it is the capital, or the process that leads to its assembly that is the more important. Thus, an organism, at whatever level, "is that in which everything is both means and ends."[32] And this is the main point I wished to make. The information contained in the genes may be necessary but not sufficient to instruct the three-dimensional development process of cells; in fact the model here suggests that interactive processes between cells and the environment in which they are embedded or intercellular exchanges must be considered. Of course, the concept of applying the I-O method to biology beggs many questions but may also provide additional tools for research on the biological discount rate and life-cycle, .

The input-output schema prescribes a steady-state situation of repetition rather than evolution and adaptation. To have reached the stage of formalizing this process, however, one should consider the search procedure for resources (energy, gradients and materials). In fact, the cells as they evolved must have been both the result and the cause for this search. Motility and transport, in the large and in the small, caused evolution and was the result of it. It is inconceivable that evolution could be carried in a replicative mode alone without the availability of environments to adapt to.

The reader will notice that I avoid the discussion of themes such as "information," "complexity", "systems," "development" as they relate to biology. The reason is the following: while the structure and performance of DNA has information content and parallels coding schemes for the precision of message communications (although not of encryptein), the

---

[32] I. Kant. See S. Korner, *Kant*, Harmondsworth, 1977.

notion has not had any known impact in the field. Information is a value-ladden concept[33] and as such cannot be of much help in evolutionary or macromolecular studies except (perhaps) *ex post.* As to the current fad with complexity, surely biological systems are complex and difficult to model; complex *models* can be worked out with computers and their evolution observed on screen. However, their relevance to actual biological processes is tenuous. If examples from physics are any guide, we observe that faced with complexity, the shedding of information to get on with what heuristically (in the beginning) might be important has provided the most powerful tools for addressing difficult problems: the many-body problems in gases and solids (superconductivity, magnetism, plasmas), and renormalization theories in particle physics. I have yet to see a computer model become superconducting (on screen) or finding an event or effect without that being *a priori* known (not impossible to be sure). This is especially true about micro- and macro-molecular physics where events themselves are not directly accessible to observation, or even possible to be observed, without destruction of the object of interest. I also avoid discussion of developmental biology (how organisms develop out of genes); while it could provide numerous ideas for "machine elements," it is outside the scope of this work.

---

[33] For the tip of the iceberg on the subject see K. Arrow, "The value and Demand of Information," Chapter.12 of his *Essays in the Theory of Risk Bearing,* Chicago, 1971: 267-278. See discussion in Appendix 4.

3. *Philosophical Issues*

The implication of the biological sketches is that to reach deeper into the origins of life we must look into the dialectical, intertemporal stochastic processes which address the interplay of capital structures and energy. The question then about the origin of capital itself becomes a fundamental one, as does the physics of its first development out of *nothing*.[34] This outlook provides a critical method for the examination of various philosophical or physical propositions. For instance, one may take exception to the melancholy view that "the more the universe seems comprehensible, the more it is also pointless."[35] The opposite might in fact be closer to reality: starting (if at all) pointless and becoming dialectically comprehensible. The Universe may yet acquire a purpose, since intelligent life may lead it in that direction. This suggestion is, of course, controversial. Recall Monod's (19..) injunctions. In his *Chance and Necessity* he demonstrates that evolution is the result of random mutations and of natural selection; yet he feels that "any mingling of knowledge with values is unlawful, forbidden." Further he claims that "the microscopic clockwork ... defies *dialectical description* and that it is thoroughly Cartesian rather than Hegelian: the cell is indeed a machine" [italics in original].[36] Of course,

---

[34] But what is nothing? In physics the closest to *nothing* is vacuum; yet even the vacuum today is populated with virtualities. Vacuum fluctuations exist and they refer to a sea of created and annihilated particles which replaces the ether of yore. See also, R. Nozick, *Philosophical Explanations*, Chapter 2, "Why is there something rather than nothing?" Harvard University Press, 1981.

[35] S. Weinberg, *The First Three Minutes*, Basic Books, 1977.

[36] "Monod finally overreached himself." See M.F. Perutz, a review of *Origins of Molecular Biology: A Tribute to Jacques Monod*, ed. A. Lwoff and A. Ullman, Acad. Press 1979 in the *Quart. Rev. Biol.*, **55**, 1980: 167. The book itself is revealing about the personality of Monod.

the prodigious stability of species and the commonality of the genetic information strongly supports the clockwork model in the short run. Yet, there is indication for an escape from the "practicoinert" and of experimentation at both the micro- and the macro-level! Errors can, and do, lead to phenotype variants so that the Cartesian vs. the Hegelian characterization becomes meaningless. Indeed, a Cartesian cellular machine would encompass no errors and, once started, would soon lead to the practicoinert state. On the contrary, the limitations in the physical ability of the cells to remain immune to noise and mutations is manifested in the discount rate at the macroscopic level. A practicoinert Cartesian system forever replicating cannot have a discount rate. At the macrolevel, the lattitude of choice in decisions is so large that, incrementally, the system is Hegelian at least in the long run. Monod's principle of invariance should be important for the short term where short term in evolutionary biology may well run into billion years.[37]

4. Conclusions

Here I have followed the philosophical critique discussed in the chapter on economics. I have viewed reductionism as both *the* inherently productive analytic instrument, when externalities can be strongly decoupled, and as an unacceptable instrument, when there is too much complexity to permit prediction and/or assessment of alternative outcomes. Mole-

---

[37] Interestingly Monod's interpretative thinking has drawn many attacks from the structuralist Marxist Althusser who criticized his inaugural address on his election to the Collège de France to his colleague A. Lwoff who according to Judson (private communication) "hated" the book *Chance and Necessity.* Indeed the book itself has political implications. One possible conclusion is that one should be generally weary of interpretative thinking even by persons who have made great contributions in their own specialty!

cular biology remains within the srictly reductionist mode of operation by decoupling externalities, while evolutionary biology provides the opposite, although complementary, outlook. The subjects discussed in this section demand a far more extensive historical and critical treatment than the limited objectives of this essay can justify. Indeed, the foundations of the biological discipline (from Darwinian evolution and its own evolution to macromeolecular chemistry) are shaken by controversies and polemics that have their source in the subjects just discussed;[38] these should preoccupy the critical reader.

In the search for the origins of capital one concludes naturally (if, perhaps, trivially) that the first structurally coherently operating capital was the replicating cell; its origins are hidden in the recesses of time and evolution. The fundamental event (invention?) was the development of a thermodynamic machine operating in a non-equilibrium environment: the replicating cell. The connection, therefore, of capital with the cell is not just a metaphor. It provides a "theory of biological form" based on principles of engineering economics, thus leading to constraints as to the alternatives that may be realized.[39] That is apparently missing from the process of evolution. Inevitably one must be guided by the principles of thermodynamic engines and their relation to their structural form, as well as at the concepts and techniques of intertem-

---

[38] For a spirited, educational, detailed and up to date review of these undercurrents the reader can profitably consult the recent book by Depew and Weber (1995) which I add as this essay is on its final edition. Unfortunately I did not have the opportunity to take advantage of its extensive scholarship (with over 700 citations) and information. The Depew and Weber book has its own proreductionist biases, however, that surface with their admiring support of the "complexity" schools of thought and its practitioners.

[39] This parallels the comments of J. Maynard Smith, in "Taking a Chance on Evolution," *The New York Review of Books*, 14 May 1992: 34 where he wonders whether there are "general principles of biological organization."

poral economics, since both have an immediate bearing on biological structures and the limitations they enforce on their performance. Any biological replicating structure *is* an engine, whether in its elemental (the cell) or in its multicellular form. Unlike thermodynamic engines, however, their biological counterparts replicate directly and, therefore, are bound to be describable through the equation of motion of economics. The appearance of the first biological cell is the original model for the production function of economics. The "stochastic" invention consisted in the formation of the capital which, in interaction with energy, is making more (and often less) of itself using specific blueprints — DNA instructions (say) or conscious planning. It is this invention in, rather than of, nature which has led to self awareness, from the *néant* to *être*. Whatever the mechanism for the initiation of the replicative system, it must operate dialectically in the field of scarcity and of capital-energy interactions; in this way "natural selection arises as near inevitability."[40] Unable to adopt any mechanism for the reconstruction of the sequence of stages that have led to the formation of the protocell, I remain sceptical with the proposals of the various schools of thought which cleave between the genotype and the phenotype; synoptic models of cellular origin are difficult to accept.

It appears that the simplistic *Central Dogma* must be integrated into a more complex and general picture of interactions and processes. Molecular assemblies are not replicating on their own. The I-O table in fact suggests that, for their replication, a complex roundaboutedness is required and that this entails capital-like structures and extended elasticities of sub-

---

[40] N. Elredge, "Information, Economics, and Evolution," *Ann. Rev. Ecol. Syst*, **17**, 1986:351-369.

stitution possibilities. The *Central Dogma* specifies one coefficient in the I-O table, while many other inputs are required to get the system activated, replicating, and mutating. The description and the "sectors" indicated are only indicative of of their complexity and they are far from being all-inclusive. There are still many details where ignorance prevails and the fact that I refer to the ATP, ADP and other such important molecules with apparent confidence does not imply that all has been said about their production, use, and localization. The opposite is truc.

As with the I-O table of economics, where the environmental conditions (the weather, the sunspots, wars, pests, etc.) as well as decision makers are nowhere accounted for, here too there are processes which are not captured by the accounting scheme of the table. Decisions, timing and sequencing of events, the environment (the properties of the heat and chemical baths), the detailed chemistry, the interaction between systems, and the transport of molecules, heat, radiation, etc. between them, are not evident. Further, nothing seems to reproduce atom for atom, position for position (mapping) a "spitting image" of the parent molecules. Rather, some sharp characteristics appear differentiated (hæmoglobin variants, for instance) because of clear and well defined DNA variations but, most of the reproduced items, are close approximations to the parents. Further, the dialectical process cannot be represented with a static I-O table. One must be able to accept that, with every cycle of reproduction, new conditions (perturbations, mutations) present themselves and are incorporated in the coefficients which then become functions of the surrounding "universe" (the externalities).

CHAPTER IV

THERMODYNAMICS ($\Theta\Delta^{cs}$), STATISTICAL MECHANICS, AND CAPITAL

1. Introduction

THIS FAR, I have referred to the relevance of thermodynamics to the theory of capital and biology only in passing; I have avoided encumbering the text or the footnotes with technicalities of the field. At this point, however, I thought it to be useful to devote a separate chapter on the critique of the subject so as to lay open to view its connection with the economic capital. I chose the route of criticism rather than that of a direct, formal, presentation for reasons that will soon become apparent.

The idea which I pursue here is the process of *reification* in thermodynamics. That is, its relationship with structures and their intertemporality. I consider these characteristics to be essential aspects of capital *and* of the discipline of thermodynamics itself. I do not question any of the formalities and findings of thermodynamics, although one can surely find faults in this or that issue; rather, I attempt to question the legitimacy of its disputable, if not shaky, foundations. Correlating the field of thermodynamics with adjacent areas (such as the theory and practice of measurement, biological work, the *utility* of work as seen in economic terms), supports my contention and I hope the reader will bear with me. The discussion suffers from the absence of solid and conclusive references to draw on, but this is what makes it enticing.

Τί δητ' ὄνομα αυτῆ θησόμεθ'
... Βούλει Νεφελοκοκκυγίαν;
'Ιοῦ, 'Ιοῦ.[1]

2. *Nephelokokkygia #1*

A famous and controversial paper on the "Reduction of the Second Law of Thermodynamics to General Mechanical Principles" was published in 1871 by R. Clausius.[2] Commenting on this paper and other related work by Boltzmann, J.C. Maxwell remarked that "the German Ikari flap their waxen wings in nephelokokkygia."[3] Klein (1970) makes the

---

[1] "And what kind of name shall we give [the town]? (815) ...What about Nephelokokkygia [*Cloud-Cuckoo-Land*]? (819), Youwho, Youwho!" (820). From the celebrated comedy of Aristophanes, *The Birds* (414 BC), being a phantasy in which an ingeneous Athenian persuades the birds to build a city in the clouds and compels the gods to accept humiliating terms. As suggested in the text that follows, the by-lines of this section have been selected to honor Maxwell's acerbic humour. Currently one uses the expression "cloud seven."

[2] *Pogg. Ann.* **142**, 1871: 433.

[3] Letter from *Maxwell-Tait Correspondence*, Tripos, December 1873, Cambridge University Library. $\Theta\Delta^{cs}$ was Maxwell's shorthand for Thermodynamics. A very attractive and informative historical review on the matter is given by M. J. Klein (1970). It can lead the interested reader to the primary sources. From this article one also gets a glimpse of the acerbic and caustic character of J.C. Maxwell. See also, Ivan Toltstoy, *James Maxwell, A Biography*, The University of Chicago Press 1982.

point — accurately I think — that Maxwell believed in the irreducibility of the second law of thermodynamics to pure dynamics. Toying with the idea of the 'demon' who could be drafted to violate the second law, Maxwell concluded that "only we can't do the work of the demons not being clever enough."[4]

The problem that Clausius believed he had resolved still haunts theoreticians despite the efforts allocated towards an acceptable resolution. The truth of the matter is that in actual work, difficult steps in "proofs" are by-passed — a method which has had no noticeable effects on the health of thermodynamics or of statistical mechanics; and it has not precluded the performance of important computations with great impact in the whole of physics. Thus, reductionism as discussed in Chapter II appears to fail in a very important area of physics. To put it in its Aristophanean form, all efforts in the direction of the resolution of the problem are still in nephelokokkygia while many more Icari are flapping their waxen wings. So many, in fact, that it has become difficult to keep track of the 'skor'[5] and even to examine critically whether the various

---

[4] On the issue of "Demon...ology" see a relatively complete bibliography "The Resource letter MD-l: Maxwell's Demon" by H.S. Leff and A.F. Rex published in the *Am. J. Phys.* 58 (3), 1990: 201-209. It consists of a much needed list of references relating to the celebrated *Demon* and to other peripheral subjects (thermodynamics, statistical mechanics, information theory, computation). The 223 annotated entries have sent many — including myself — to the library seeking missed pieces not previously consulted. Others, however, may soon be pointing to incomplete coverage since it appears impossible to confine the *Demon* to a narrow intellectual territory. *Maxwell's Demon — Entropy, Information, Computing,* Princeton University Press 1990 which followed, contains an overview of the field, twenty-five reprints of original articles and a chronological and an alphabetical bibliography.

[5] Τό σκῶρ, τοῦ σκατός; see Liddell and Scott, *Greek-English Lexicon,* Oxford 1968. With 500 hundred citations in Wehrl (1978), as many in Penrose (1979), 251 in Leff & Rex (1990), the pun is irresistible.

contributions address the same problem, variants thereof, or some of the offshoots.

One can discern two schools of thought which have developed. The one belongs to the antireductionist camp which gives primordial, axiomatic content to the second law; the other, belongs in the reductionist camp which tries to prove its content on the basis of "first principles" — i.e., classical and quantum mechanics. A prominent representative of the antireductionist, idealist, school is Max Planck who, aware of the difficulties of the microscopic approach, settled for the axiomatics of the "constrained irreversibility" in his influential and multiply revised and reprinted *Treatise on Thermodynamics* (1945). On p. 89 he adopts Wilhelm Ostwald's formulation to the effect that

> it is impossible to construct an engine which will work in a complete cycle, and produce no effect on its environment other than the raising of a weight and the cooling of a heat-reservoir (an axiom),

and then proceeds to refer to the "most profitable engine" that would otherwise be possible; should it be feasible to construct such an engine, he argues, "it would possess for man the essential advantage of perpetual motion, the supply of work without cost" — even the cost of the initial capital disappears in an infinite regress! Note that Planck (*ibid.* p. 106) also states that the "gist of the second law has nothing to do with experiment...," that the law is valid "irrespectively of whether human beings exist on earth or not" and that "the limitations of the law if any, must lie in the same province as its essential idea, in the observed nature, and not in the observer."[6] Further, he rejects the idea that the "validity of the second law depends on the skill of the physicist or chemist in observing or experimenting" (no demons admitted), and he states emphatically that the "law once discovered must receive recognition of its independence, at least in so far as Natural Law can be said to exist independent of Mind." He adds "that man's experience is called upon in the deduction of the law is of no consequence." At the end of the particular chapter he adds that "presumably the time will come when the principle of the increase of entropy will be presented without any connection with experiment. Some metaphysicians may even put it forward as being *a priori valid.*" I retain the quotations from Planck in the main text because of their provocative nature, the prestige of their author, and because generations of physicists and engineers have been nurtured by Planck's work — a fact attested by its numerous editions.

The philosophical statements, of course, in no way affect the phenomenology of thermodynamics in the working text, although they have influenced historians of science and philosophers, specialists often unfamiliar with the structural and functional parts of the discipline. In any case both the

[6] *ibid.* p. 105-107. The question whether truth is created or discovered is, of course an old one. Planck's position is that truth is there to be discovered while Nietzsche, for instance, takes the opposite view: "'Truth' is ... not something there, that might be found or discovered — but something that must be created and that gives a name to a process, or rather to a will to overcome that has in itself no end — introducing truth as a *process in infinitum*, an active determining — not a becoming conscious of something that in itself is firm and determined." F. Nietzsche, *The Will to Power*, transl. by W. Kaufmann aand R.J. Hollingdale, Vintage Press, 1968: 522. This echoes contemporary concerns. Nietzsche however, did not suggest that the laws of nature themselves are created; but that they are dependent on the development and evolution of novel cognitive processes for their recognition: creating truths which in turn constrain creations.

axiomatics and the philosophy of Planck leave much to be desired but they touch upon issues which still plague the foundations of physics as is evidenced by the raging contemporary controversy on the *anthropic principle*. But take note of the words that Planck used in the formulation of his axiom: *impossible, construct, engine, complete, produce...*; and those used in the commentary: *profitable, cost, etc.* This is not language fit for axioms. Indeed, the axiomatics of thermodynamics needed cleansing and this was provided with some elegance, finality, and completeness (in the axiomatic sense) by Carathéodory (1909). It was an attempt to expurgate thermodynamics and the Second Law from the Carnot engines and cycles thus introducing a measure of abstraction by converting well understood physical qualities (adiabatic, say) into protophysical looking concepts demanding of no further definitions. The language is terse; the words imply concepts rather than anthropomorphic characteristics. Pedagogically sterile, it is not material a design engineer will use; in fact axioms must be pragmatically useful; as Gödel remarked: "even disregarding the intrinsic necessity of some axiom, and even in case it has no intrinsic necessity at all, a possible decision on its truth is possible also in another way, namely inductively by studying its success."[7] Success means fruitfulness in consequences, in particular "verifiable" consequences. On the other hand heuristic stimuli should not be overlooked. Einstein, for instance, despaired in trying to discover a universal formal principle in the understanding of light propagation. But he writes that "the example I saw before me was thermodynamics. The general principle was there given in the theorem: the laws of nature

[7] K. Gödel, "Russell's mathematical logic, and what is Cantor's continuum problem?" in *Philosophy and Mathematics*, P. Benacerraf and H. Putnam (eds.), Prentice Hall, 1964: 211-232, 258-273.

are such that it is impossible to construct a *perpetuum mobile* (of the first or the second kind)..."[8] This heuristic led to the formulation of the principle of relativity that stems from the equivalence of observers in relative uniform motion.

If Planck used the language of economics to describe the implications of the *Second Law*, economists have not stopped having a fascination with thermodynamics. But its formalism, not its substance, proved useful to mathematical economics. Georgescu-Rœgen (1971: 169), an economist, who tried to deal with the substance did not get very far. He notes succintly, however, that "of all physical concepts only those of thermodynamics have their roots in economic value, and hence could make absolutely no sense to a nonanthropomorphic intellect!" But he contradicts himself when he asserts that "the Entropy Law represents another irreducible law of nature alongside Newton's law of inertia." Here we have a case of a non-physicist who felt the need — if not the obligation — to dwell upon the foundations of thermodynamics because the economic process is necessarily one dependent on energy transactions. The argument goes to show that there is a need for better thermodynamic texts with critical and accurate assessments about the state of the art in the field.[9] Having discussed the ambiguities that infest thermodynamics by using samples from statements and opinions of various scientists, I now venture some assessments:

---

[8] "Autobiographical Notes," in *Albert Einstein: Philosopher Scientist*, P.A. Schlipp, ed., Library of Living Philosophers, Evanston, Ill. 1949: 53. The German refers to "satz" that does not translate monovalently as "theorem"; Einstein may have meant "statement," "tenet."

[9] In this direction useful review articles by Wehrl (1978) and Penrose (1979) can be consulted.

The significant contribution of Carathéodory has been the *derivation* of the expression for entropy from a general set of postulates mentioned earlier thus disassociating the logical foundations of thermodynamics from the specific realization of a process, namely the Carnot cycle.[10] This emancipation from the mechanical image has regularized the formal structure of thermodynamics but it is a dead end for it does not open any new predictive or heuristic avenues. As it stands, the second law and the concommitant measure of entropy are describing a macroscopic empirical phenomenon with magesterial finality. As Einstein noted, "within the framework of applicability of its basic concepts, it [thermodynamics] will never be overthrown."[11] To my judgement, however, the significance of the second law rests in the incisive way with which it summarizes a diversity of microscopic events. We can only speculate that the law incorporates quite economically fundamental processes which are subject to microphysical specifications of the region in question. After all, this is why thermodynamics is such a poweful scientific tool. It goes beyond the physical detail to an abstraction which covers all cases of energy interactions and energy events. Interestingly, it covers not only the smoothed average performance of systems in their macroscopic phenomenology but also reaches one layer below, in order to account for fluctuations. It intimates, yet never really depends, on the microscopicity of physical events; and yet no conjectured mi-

---

[10] On the other hand Eigen (1971) in an important paper on the self-organization of matter erroneously states "that with entropy a new 'concept' was introduced becomes obvious with the axiomatic foundations of thermodynamics given by Carathéodory and others." Entropy is a definition not a concept.

[11] "Autobiographical Notes," *loc. cit.* For the "Thermodynamics in Einstein's Thought," see M. J. Klein, *Science*, **157**, 1967: 509-516; a valuable and perceptive historical analysis that traces Einstein's search for fundamentals.

croscopic event can be "observed" and interpreted without support from the phenomenology of thermodynamics. The second law is accepted as an indispensable statement of phenomenological macrophysics which guides the design and the performance of equipment and devices. Yet, in its abstraction, this most realistic and pragmatic of disciplines has become disembodied — it has been left without a reference to the structural boundaries within which its prescriptions operate. *How* this works out is the fundamental question. But the question is tricky and the ground slippery. We may conceive of a research program addressing the legitimation of thermodynamics as a game whereupon every time a researcher lands on a demon's square he is made to pay a price or return to square #1: the winner must reach the end without invoking demons of any disguises; and referees must be able to see through them. The trouble begins with the reductionist inquiry relating to the microscopic origin of entropy. This is the subject of Nephelokokkygia #2, the section following a small digression.

Wasps (Σφῆκες) [12]

Counterpoint. The popular fascination with thermodynamics and the measure of entropy is itself an absorbing subject worthy of psychological interpretations. A list of references in which *entropy* has been used, mentioned in passing, or in substantive passages not directly connected with the technicalities of the discipline, can run into a small book. The range of these references can be reckoned with by considering that

---

[12] A play by Aristophanes where idle older Athenians suffer from jury mania. Disguised as Wasps with ... stings, they make up the chorus.

thermodynamics in its heyday had inspired the Saint-Simonian school of planners — anathema to Hayek (1955)— that in its reincarnation during the energy crisis of the seventies it had become the focal point of energy conservation, and that it has inspired some admirable musings in E. Arsan's erotic novel *Emmanuelle*.[13] Indeed, the "great" Carnot[14] himself was one of the most ardent supporters of Saint-Simon. Arago wrote that "Carnot always discoursed ... on the political organization of society precisely as he speaks in his works of a machine"! One can then imagine the impact of the subsequent discovery by his son Sadi of the (thermodynamic) cycle that bears his name and, in particular, the cycle with which the ultimate technical efficiency for conversion of heat to directed motion was determined. Thermodynamics has acquired a degree of notoriety of late; Christopher Lehman-Haupt of the New York Times in his "Thermodynamic Socialism" (1976: 31 May), a title echoing the Saint-Simonian "scientific socialism," was humoring Barry Commoner's *The Poverty of Power. Energy and the Economic Crisis.* Commoner felt the need to prepare a popular presentation of thermodynamics in support of his policy views on the then trendy energy conservation. Although admittedly the technical presentation is faultless and is accompanied by *New Yorker* style illustrations appropriately describing the evolution of physical events from coherent to incoherent (wiggly), the suggestions for conservation along the 2nd Law of $\Theta\Delta^{cs}$

---

13 Grove Press 1971: 161.

14 Lazare, the father of Sadi and Hippolyte published the *Principes Fondamentaux de l' Équilibre du Mouvement* (1803) which was followed in 1824 by Sadi's celebrated *Reflections sur la Puissance Motrice du Feu et sur les Machines propres à Développer cette Puissance,* Bachelier, 1824. Hippolyte served as minister of public instruction in the provisional government of 1848 and was also a "polytéchnicien" and a Saint-Simonian; he attempted to create the first School of Public Administration. For a perceptive and critical study of the period refer to Hayek (1955).

leaves much to be desired. Abstention from energy consumption can be accomplished with various economic and technical means without immediate or even long term recourse to the 2nd Law. But it supports the contention that "das Charakteristikum des Weltanschaung des Ingenieurs ist die Energetische Weltanschaung."[15] Clearly, the exhortations of the seventies on energy conservation do fit quite nicely with Hayek's (1955) analysis of the psychology of the engineer who "feels" that "irrational economic forces prevent him from achieving the ideal." We should not despair, however, as the Energy Crisis ("Entropy Crisis" to some!) has come to rescue thermodynamics from intellectual extinction. On the other hand and at another level of discourse C. P. Snow in his "Two-Cultures" theme (so popular among the upper-mobile scientists of the sixties) used the knowledge of the 2nd Law as a measure of literacy in the sciences among non-scientists; presumably all scientists understand the 2nd Law and, not unlike J.C. Maxwell, can also recite their classics. Recall too that in an unusual editorial the *New York Times* (29 May 1979) lyrically extolled the thermodynamic *Loophole* [sic] which, as the editorial suggested, had resulted from the recent Prigogine findings reported a few days earlier in the same newspaper by Brown.

Tiptoe around, traveller, lest ye rouse
the hornets resting on this [my] tomb[16]

[15] L. Brinkmann, *Der Ingenieur*, Frankfurt, 1908: 16, quoted in Hayek (1955: 220).

[16] "ἠρέμα δεῖ παράμειψον, ὁδοιπόρε, μὴ ποτὲ τοῦδε κινήσῃς τύμβῳ σφῆκας ἐφεζόμενους," On Archilochos's tomb c. 700 B.C. Palatine Anthology: Gætulicus.

Ἀρ' ἐστίν αὔτηγί Νεφελοκοκκυγία;[17]

3. Nephelokokkygia #2

It is generally agreed that the objective of statistical mechanics is to derive the theorems of phenomenological thermodynamics as they arise from microscopic many-body classical or quantum particle dynamics. Yet, whether the reductionist problem has been solved, is about to be solved, or, for that matter, is well posed at all, remains strictly a matter of opinion. In the introduction of a book by Krylof (1979), Sinai *et al.* state categorically that "the problem of establishing the connection between statistics and mechanics should be regarded as absolutely unsolved." Also A. Wightman writes that "one of the great problems of equilibrium statistical mechanics is to show how the laws of thermodynamics arise from microscopic dynamics" (Israel 1979). Those supporting this position openly, are few and the confusion is quite great especially among generalists. The misconception expressed by Jacob (1977: 162) that "thermodynamics and mechanics were unified through statistical mechanics as were optics and electromagnetism through Maxwell's theory of the magnetic [sic] field..." is widespread. Indeed, optics and electromagnetism were shown to be one and the same thing describable through one set of equations — Maxwell's. This is not the case with thermodynamics and mechanics (classical or quantum). Among reductionists, thermodynamics is thought to be just another, more aggregate description of mechanics; so it is not a matter of unification but rather of a missing proof.

[17] (821). "could this be the real Cloud-Cuckoo-Land?"

The corresponding problem in electromagnetism, which consists of the derivation of the macroscopic phenomenological equations starting from microscopic equations of motion, is entirely dependent on the legitimacy of thermodynamics as an offshoot of mechanics. All this goes to show how important it is to come to terms with the foundations of thermodynamics, i.e. to address and test the reductionist program.

The reductionist program consists of loops in research which presumably converge to the "truth." This program is sketched in figure IV.1 and sums up the methodology reasonably well. Not all researchers have followed the program consciously or purposefully and leaps of imagination have often proven more fruitful. All theoretical constructs originate from the images of macroscopic phenomena while all microscopic phenomena can only be inferred — but never directly observed — through macroscopic effects. The process for the development of these constructs is a spiralling one; a never ending search, interrogation, and learning, while novel theories and new theoretical and experimental techniques are developed to address inconsistencies. We have been led to believe that the spectacular successes of mathematical physics over the last century are an indication that the idealist school (neoplatonism) was the guide towards unraveling nature's ways. This entails the search for "guidelines" that constrain the choices for equations of motion (gauges, renormalization, etc.) i.e. the adoption of the "synoptic" view of theories and theoretical entities. However, even the best (most successful) of theories will not survive scrutiny in this respect for they not only leave some serious foundational problems unresolved and mysteriously *ad hoc* but they also

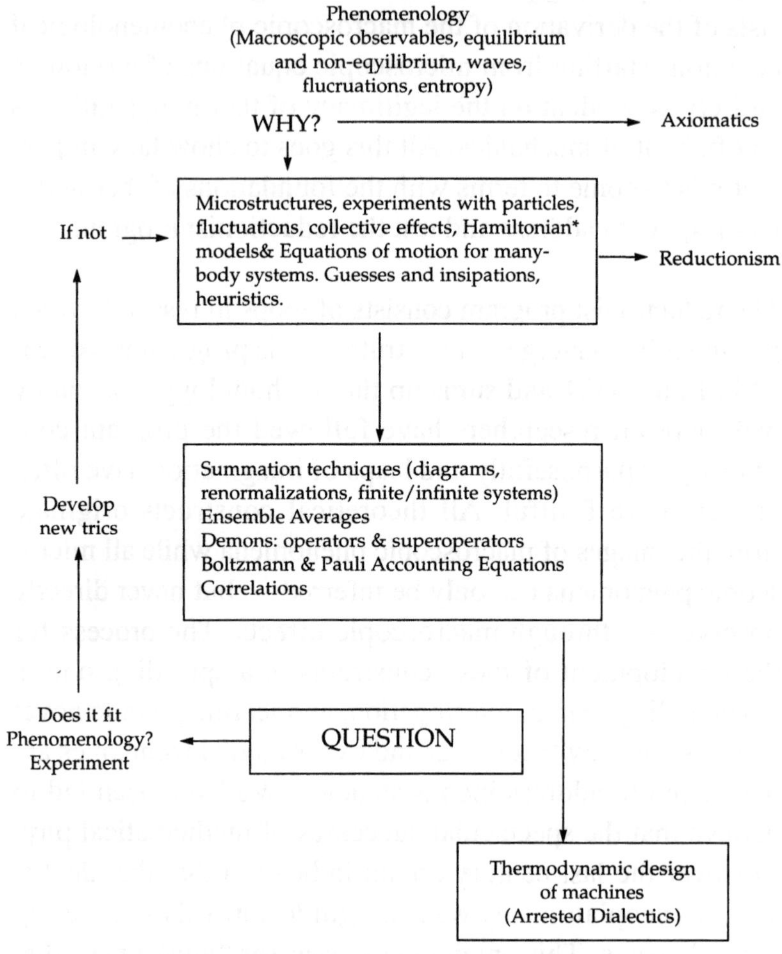

Fig. IV-1 A Reductionist, Meliorative, Program for the Development of Theoretical-Physical Constructs

***The Hamiltonian is the key and is invariably guessed especially as concerns its interaction terms. The <u>mechanical</u> terms — kinetic energy — are always the same**

appear to be valid for limited regions.[18] So that the incremental approach depicted in Fig. IV.1 is, perhaps, closer to what one does in science rather than what one says or imagines. This incremental approach an "endless chain of give and take, the facts suggesting new analytic instruments (theories) and these in turn carrying us towards the recognition of new facts," is according to Schumpeter equally valid whether the object is history or "to 'rationalize' the Schrödinger equation, though in any particular instance the task of fact finding or the task of analyzing may so dominate the others as to remove it out of sight."[19]

In the theoretical plane the effort consists in the development of macroscopic equations from classical (CM) or quantum mechanical (QM) *microscopic* equations of motion. There is a blatant circularity here since microscopic equations of motion are invariably inferred from, or guessed to fit, macroscopic experiences. Those familiar with the process are only painfully aware that a moment comes in the calculations where the principle of satificing takes over and — with a magic wand — produces the equations just one layer of calculability below what is provided in thermodynamics: the (irreversible) *Boltzmann Equation* for transport of matter and the *Pauli Master Equation* for transitions.[20] The first equation

---

[18] Cao and Schweber (1993: 74) call this "localism."

[19] Joseph A. Schumpeter, "Science and Ideology," *Am. Econ. Rev.*, XXXIX, 1949: 350-351. Comparing this statement with the one on the independence of economics (p. 16) it is apparent that Schumpeter here acknowledges a unity among disciplines of the theory/experiment nexus.

[20] For paedagogical reasons I have found that the work and research on Plasma Physics over the last 30 years has done much to clear the field because plasmas are relatively straightforward systems in terms of interactions. For a clear and often original exposition see Yu.L. Klimontovich, *The Statistical Theory of Non-Equilibrium Processes in a Plasma*, transl. H.S. Massey, The MIT Press 1967, where the extensive use of the *second quantization te*chnique provides an intuitively pleasant exposition. The

refers to the (surmised) distribution $f(p,q,\mathrm{t})$ of particles in phase space (momentum, position, time) while the second refers to the (surmised) rates of change in energy states in a system. Both equations are intuitively secure and they describe events one layer of sophistication below the phenomenology of thermodynamics by introducing distribution functions and events (processes). They allow a semblance of the particulate nature of a fluid where thermodynamics allows only for a fluid continuum. The words used are thus deceiving as the equations have yet to be derived from fundamentals. I do not propose to develop these equations here but they are used mindlessly and/or mechanically in thousands of papers while the primary sources of the technique remain difficult and often obscure. Decades have been devoted to exactly this enterprise; to derive, that is, these powerful equations starting from "elementary" building blocks and principles which consist of the Hamiltonian[21] of the underlying microworld, the equations of motion — classical or quantum — etc. As these two equations are now presented they are *stricto sensu* accounting equations which, on "summation," deliver macroscopic descriptions. For example, the macroscopic density of a system $\varrho(q,t)$ is the integral of the — still phenomenological — distribution function $f(p,q,\mathrm{t})$ over momentum space:

$$\varrho(q,t) = \int f(p,q,t)dp. \tag{4.1}$$

more advanced work should be sought under the "many-body theories," "Thermodynamic Green's Functions," etc.

[21] The "Hamiltonian" H, is the total energy of a system (kinetic + potential + electromagnetic). From it the "canonical" equations of motion can be derived using the expressions: $\dot{p} = -\partial H/\partial q;\ \dot{q} = \partial H/\partial p$.

f itself is the one-particle correlation function or, the "ensemble" average denoted by $\langle\ \rangle$ or $\bar{\ }$ (on which more below) of the 'microscopic' sum of the distributions

$$f(p,q,t) = \langle N(p,q,t)\rangle = \langle \sum_{i}^{N} \delta(p_i(t) - p)\delta(q_i(t) - q)\rangle \quad (4.2)$$

in phase space.[22] Whether in classical or quantum mechanics, one *pretends* that the microscopic structure of the physical world can be described symbolically by this innocuous-looking point-particle expression yet it is problematic for many and important reasons. It is, nevertheless the expression universally used under one or another guise to *represent* the contents of a finite system. A finite system that contains N particles separates the universe in two parts which are of necessity interacting, however mildly. N changes and fluctuates as particles are created and annihilated, chemical species are synthesized and lysed, atoms are ionized and deionized, and all these move in and out of the 'boundaries' of the system under investigation. It is in this terribly complex event space that we attempt to make some sense.

Rather than embarking on the substantive aspects of the distribution function, I will provide a few straightforward formalistic manipulations. First I will compute the time rate of change of $N(p,q,t)$:

---

[22] The celebrated "delta function" $\delta(x-x')$ was introduced in mathematical physics and legitimized in mathematics by Schwartz. It simply expresses that when you pin a thread to a point and rotate it around in space it will register an event when it hits a "point particle." The density of particles is clearly infinite at that point since we have 1 divided by $0m^3$. This is the definition of the delta function: an anomaly in physics where one needs smooth functions. However the integral of the function is finite and this saves the situation: $\int\delta(x - x')dx = 1$.

$$\frac{dN}{dt} = \frac{\partial N}{\partial t} + \frac{\partial N \partial p}{\partial p \partial t} + \frac{\partial N \partial q}{\partial q \partial t}, \quad (4.2) \qquad (4.3)$$

and then substitute the momentum-position rates of change with their values in terms of the Hamiltonian H (see ftn. 19):

$$\frac{dN}{dt} = \frac{\partial N}{\partial t} - \frac{\partial N \partial H}{\partial p \partial q} + \frac{\partial N \partial H}{\partial q \partial p}. \qquad (4.4)$$

This is a very important result-generating expression. For the moment we can get a first easy clue as to what is going on by specializing the Hamiltonian H to an electron plasma where $H \doteq p^2/2m + e\varphi(\mathbf{q},t)$. With N of Eq. 4.2, the Poisson equation becomes

$$\nabla^2 \varphi(q,t) = \pm 4\pi e N(p,q,t) \text{ i.e. } \varphi(q,t) = e\int \frac{N(p,q',t)}{|q-q'|} dq'.$$

Assume now that there is no net particle creation, annihilation, entry or exit from the system. Then $dN/dt = 0$ for the left-hand side of 4.4 which now takes the form:

$$\begin{aligned} \frac{dN}{dt} = 0 &= \frac{\partial N}{\partial t} - \frac{\partial N}{\partial q}\frac{p}{m} + e\frac{\partial N}{\partial p}\frac{\partial}{\partial q}\int \frac{N(p,q',t)}{|q-q'|} dq' \\ &= \frac{\partial N}{\partial t} - \frac{\partial N}{\partial q}\frac{p}{m} + e\frac{\partial}{\partial p}\frac{\partial}{\partial q}\int \frac{N(p,q,t)N(p,q',t)}{|q-q'|} dq'. \end{aligned} \qquad (4.5)$$

I would not have gone through the trouble of "deriving" the above equation[23] had it not been for the purpose of making a point. We started with a "one-particle" distribution function N and its time evolution only to find that we are only starting a chain where N is connected with the "two-particle" distribution function NN — a process which can be continued by writing the equation of motion of NN in terms of NNN and so on until we exhaust all the N particles. Then, in each step, an averaging procedure must be applied and terms such as $\langle N \rangle$, $\langle NN \rangle$, $\langle NNN \rangle$,..., sorted out. The above expression and the chain that follows is quite general and applicable to both classical and quantum systems provided the appropriate definitions of N are introduced [in the quantum case (in 2nd Quantization) $N = \psi(q,t)\psi^{\dagger}(q,t)$]. The information for all this is enormous; to handle it in real time we resort to the process of jettisoning — at least in the theoretical work. We do believe, however, that what is observed in the thermodynamic world is an average (smoothed in space and time) behaviour of these particles. Invariably the behaviour is what we call classical; occasionally, as in superfluidity and superconductivity, we also observe quantum effects in the large.

On the two equations referred to earlier (Boltzmann, Pauli) and their many variants, rest the modern analyses of the theory of bulk matter. Under different guises, simplifications and calculations, they have truly become the workhorses of modern classical and quantum physics and of some engineering too. They have no explanatory value and often pose insuperable problems of interpretation and utilization. De-

[23] Formalistically this simple looking but in fact quite complex equation is very close to the celebrated Landau-Vlassov expression from which derive the ideas of plasma oscillations, and Landau damping.

spite their strong flavor of atomicity, both equations are within the classical thermodynamic tradition and, as such, cannot on their own, 'legitimize' thermodynamics; they simply extend its reach. Their own legitimacy, i.e. their derivability from first principles, is itself in question and this has led to the use of Nephelokokkygia #2 for the title of this section.

I, now, sketch the steps which specialists adopt on their way to legitimizing thermodynamics. They utilize statistical techniques of averaging and the more abstract, if very controversial, idea of the *ensemble*, an imagery first introduced by Gibbs. It requires a mental abstraction which is beautiful to contemplate but difficult to believe. Instead of dealing with the presumed microscopic structure of assemblies of weakly interacting particle systems (atoms, molecules, photons, etc) Gibbs proposed the study of N identical *systems* (not particles), all mental copies of the one under examination — the ensemble. Schrödinger (1952) in his celebrated *Statistical Thermodynamics* asks "what on earth could it mean, physically to distribute a given amount of energy E over these N mental copies?" and, compounding the puzzle, suggests that the systems are "in 'weak interaction' with each other, but isolated from *the rest* of the world." It is as if though one of them is "in a peculiar kind of 'heat-bath' which consists of the N-1 others." One can also interpret the ensemble as referring to the sequence of states that a system can take. But in fact there cannot be any realism in the *ensemble*. The ensemble is a trick to reach the infinity needed to have equations that make sense. But the notion stretches the nerves of all researchers and leaves them helpless, if not cynical. Dirac has characterized the ensembles as "theoretically useful constructs," while Deutsch (1985) believes that the ensemble provides an "infinity" which is mathematically desirable if not necessary. Lebowitz and Penrose (1972) further assess the situation stating that "the distinction between microscopic and macroscopic observables which appears essential to any complete theory of irreversibility and kinetic equations can be formulated precisely for infinite systems." A clarification on these issues is now due to an important advance in the field of non-equilibrium statistical mechanics that is contained in the so called "KMS-condition." The legitimation of the Gibbs ensembles for finite systems in equilibrium comes from the following argument.

In particle physics events occur in a "vacuum."[24] By contrast, in thermodynamic systems events occur in a "plenum" filled with elementary states of thermodynamic equilibrium. The equilibrium state is stationary so that its distribution function (or, equivalently, its quantum mechanical density matrix) should be a function of the constants of the motion some of which are known a priori: energy (H) and particle number (N). The system has N particles in a "box" $L^3 = \mathcal{V}$ while the total energy is E. Only when N, $\mathcal{V}$, E go to infinity while the ratios $N/\mathcal{V}$, $E/\mathcal{V}$ remain finite do thermodynamic quantities acquire meaning and then, by basing the theory on the algebra $\mathcal{A}$ of local observables, the box becomes dispensable. In engineering one operates on the basis of $\mathcal{A}$, the elements of which are the "intensive quantities".[25] The Gibbs states defined in Appendix 1, namely the *Gibbs Ansatz,*

---

[24] In quotation marks because by today's physics the vaccum is a very lively container of events.

[25] Intensive variables are size-independent (for instance, temperature T, pressure P); volume V and entropy S are extensive variables i.e. proportional to size.

$$f_N(p,q) = Z(\beta)^{-1} e^{-\beta H(p,q)}; \quad Z(\beta) = \int d\Gamma e^{-\beta H(p,q)}$$

have a property that was first described by Kubo[26] (1957) and subsequently and independently applied by Martin and Schwinger (1959) in the development of their thermodynamic Green's functions—hence the KMS designation. The gist of the idea rests on the combination of the time evolution of an observable with the Gibbs ansatz. Indeed the time translate of an observable A is given by $a_t(A) = e^{iHt}Ae^{-iHt}$ where H is the Hamiltonian so that on combination with an average taking with $e^{-\beta(H-\mu N)}$ we end up with a complex variable $z = t + i\gamma \rightarrow t + i\beta$ so that we deal with exponents in the form $e^{\pm iHz}$ that lead to quantities that survive the thermodynamic limit of $\mathcal{V} \rightarrow \infty$. This development is mathematically very difficult to follow and cannot be verbalized. The best I can do is to accept it on faith, reproduce the concluding statement of Haag (1992: 207), and leave it at that: "the advantage of the KMS-condition over the Gibbs prescription [ansatz] is that it carries over to the infinitely extended medium. There, $\omega_{\mu,\beta}$ can no longer be described by a density operator...but as a state over an abstract C*-algebra ... This eliminates the box, the boundary conditions, the consideration of the thermodynamic limit ..." One, then, may conclude that the ensemble proved to be a safe expedient to reach an objective that the KMS now legitimizes. As with so

---

26 See ftn. 1 of Introduction. Kubo's paper may be thought of as being a generalization of H. Nyquist's theorem, a classic in electrical engineering, which relates the impedance of a circuit to the thermal noise in its resistive parts: "Thermal Agitation of Electron Charge in Conductors," *Phys. Rev.*, **32**, 1928: 110-113.

many modern developments in mathematics (the Fermat Theorem comes immediately to mind) and mathematical physics (say, the KMS-condition) the infrastructure and roundaboutedness needed to prove heuristics is forbidding and accessible to very few; but this should not lead one to dismiss heuristics altogether!

The interested reader will now have to consult textbooks and Appendix 1 where I sketch, without any pretense at rigour, the sequence of steps which lead to the development of *all* the formularium of thermodynamics beginning from statistical mechanics basics. The assumption, then, is that we know how to compute statistically.[27] To go deeper, however, we must deal with the complete panoply of the many-body theory starting with the microscopic equations of motion which in their innocent looking form deal with each particle in a large or even infinite system, and proceed to the justifications which permit the use of the symbol $\langle\,\rangle$. However, the microscopic equations of motion, derived simply from the knowledge of the Hamiltonian of the system (kinetic + potential energy) are equations reversible in time: changing t to -t, the manifold backcasts![28] The issue is that experience indicates that events move only in one direction — they are irreversible — while *all* the basic equations of motion in

[27] The procedure in statistical mechanics is first to identify a (the) system in interaction with a "vast" thermostatic bath (a bath whose existence is not questioned). The time-independent distribution $f_N(\mathbf{p},\mathbf{q})$, is called microcanonical when $f_N(\mathbf{p},\mathbf{q}) \propto \delta(H(\mathbf{p},\mathbf{q}) - E_0)$ where δ is the Dirac delta function, macrocanonical when $f_N \propto V_{therm}(H(\mathbf{p},\mathbf{q}) - E_0)$. V_{therm} is a function of the heat bath; grand canonical ensembles refer to systems which, in addition, exchange particles.

[28] Time-reversal invariance implies that equations remain unchanged when t is substituted for -t, i.e. if $\partial a / \partial t = R$, then also $\partial a / \partial(-t) = R$. Here R is a dynamical rule for the variable a. Irreversible rules do not satisfy this time symmetry.

physics are invariant under time-reversal. How can this be is the central problem of statistical mechanics and, more generally, of physics. Irreversibility has been recognized sincethe dawn of history — "Rivers run backwards" ["Ανω Ποταμῶν] was an exclamation of disbelief.[29] Yet the formal description of this phenomenon escapes physicists who clearly resort to rhetoric. The resolution of this difficulty is the fundamental object of statistical mechanics. There is no way that the fundamental equations of motion can be made to (honestly) exhibit irreversibility in the macrophysical world except by using "demonian" tricks. In the face of this position I quote from a recent review by Prigogine and Grecos (1979): "we believe," these authors remark, "that every dynamical theory of irreversibility should ultimately lead to a microscopic interpretation of the second law of thermodynamics and the concept of entropy ... Irreversible processes cannot be considered as arising from some "accident," some improbable initial condition, or our ignorance of the 'exact' dynamical state" (*ibid.* p. 311). In response to this logic Prigogine and his school have attempted to develop what appears to be an ansatz through the utilization of projection operators which ensure an *ab initio* microscopic irreversibility. Mathematically, an equation of motion which is written as reversible cannot be converted into an irreversible one except by fiat i.e. by invoking some intuitive process or approximation. One gets the impression that the various au

29 ""Ανω ποταμῶν ἱερῶν χωροῦσι παγαί" Euripides, Medea 413. And Julian (A.D. 331-362) in his address to "the uneducated cynic" laments exclaiming ""Ανω ποταμῶν, τοῦτο δὲ τῆς παροιμίας," (Up flowing rivers [i.e. ridiculous], as the saying goes), *The Works of the Emperor Julian*, Vol II, The Loeb Classical Library. But for William H. Calvin's *The River that Flows Uphill —A Journey from the Big Bang to the Big Brain*, Macmillan, New York, 1986, I was tempted to use this as the title of this book.

thors have developed necessary guidelines which arbitrarily select conditions for irreversibility.[30]

Νῦν τε λόγων πτερῶσε [31]

4. Intermezzo

I have been critical of the state of affairs in the efforts to legitimize phenomenological thermodynamics without, however, being able to suggest alternatives; yet, I doubt that alternatives to the work of Gibbs and its KMS extension will be forthcoming soon. The reason is that the problems seem to be beyond the current methodological-mathematical techniques although they may be described in words. Let us see why this is so.

Science must proceed with incomplete logical steps — often justifiable *ex post* — along with imaginative thinking and questioning. Note, for instance, that all theoretical work in physics begins by writing down a (model) Hamiltonian — the energy of a system under investigation — from which the equations of motion can be derived whether in their classical or quantum form. Yet there is never a questioning as to how the system came into being in the first place (see Figure IV.2); a "let there be" bridges the gap between nothingness and being. Consider the problem of the nucleation of mat-

---

[30] Prigogine (1977) constructs a "unitary transformation" and a controversial "microscopic entropy operator" with serious philosophical antireductionist implications; with these he seeks a "far reaching modification of dynamics," (1980). Zubarev (1974), on the other hand, builds advanced and retarded solutions into his equations so that the advanced ones have an edge over the retarded ones (technically, he choses the right poles). And so on.
[31] *ibid.* 1437: Now then, with words I'll fit you with feathers.

ter.[32] It is generally taken for granted by the theoreticians although measures for the cohesive energy and of the energy required for the addition and the removal of particles is available.[33] Nevertheless, it is quite difficult to develop theories that encompass extremes of physical behaviour and physical structure and that, in addition, include state and phase transformations.

Formally, to describe the evolution of the construction of a system, an algorithmic procedure would involve an aleatory sequence of incremental Hamiltonians of the form

$$H_n = \lambda U H_{n-1} U^{-1}$$

with the n-th Hamiltonian H_n derivable from the (n-1)th, through the unitary operators U. I have borrowed this idea from Dyson (1979) who, in another context, has suggested such an algorithm as a scaling, adaptive device, in the description of the evolution of life towards a different thermodynamic environment. The idea is seminal in that it describes

---

[32] The reader is referred to Lieb's 1976 review on the studies for the *stability of matter* which are important and, in a sense, striking. Schrödinger's equation is 'silent' on this score although when results are worked out, matter does not collapse.

[33] F. Seitz,1940: 345; and 313-314 on Koopmans' 1933 theorem. That the Hamiltonian changes during the interaction is usually ignored. There has been one instance where this has been accounted for and it is to be found in Koopmans' theorem on the energy needed to remove an electron from a solid. In the application, however, the wave functions ψ are considered to be the same before as they are after the interaction so that the energy difference between the two states takes the form $\int \psi_j^*(H'-H)\,\psi_j$ where H and H' are the *before and after* Hamiltonians respectively. In an endeavor to describe the formation of molecules, especially the large and complex ones, we must be dealing with a step-wise change in the time variable of the Hamiltonian: H(t) = H(t-1) + ... This is the procedure to follow in the construction of structures which we could call generic capital. The interaction of this capital with energy provides both the elements of reproduction and the system to handle non-equilibrium processes.

structural change without, however, providing the detailed procedural mechanisms for this change. The setting of the problem brings forth the whole problematics of condensation and of the stability of matter.

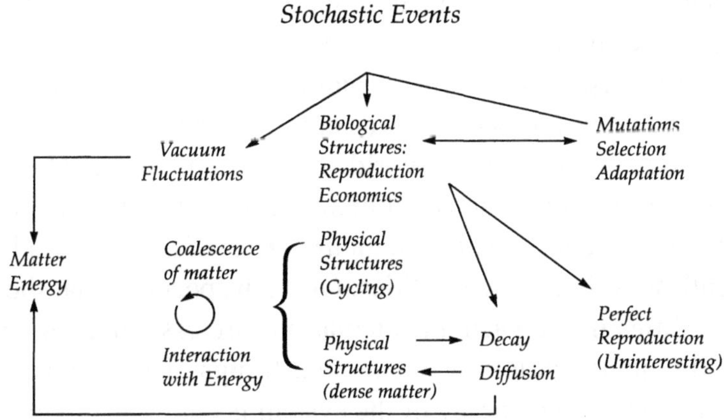

Figure IV-2. Evolution of Structures

A plausible evolutionary sequence of events from a matter-energy environment beginning to what is observed in the biological and economic replicating structures. A reductionist attempt!

In a classical world, particles cannot coalesce to form molecules, macromolecules, cells, solids, etc.; an electron orbiting a nucleus radiates energy and soon winds up on the nucleus. The stability of matter is an inherently quantum mechanical manifestation. Curiously, such an important topic is not stressed in textbooks. Schrödinger (1945: 50) was one of the first to recognize this in his essay *What is Life?* as regards the stability of molecules ("the molecule will of necessity have a certain stability. ... It will be observed how intimately this fact is linked with the very basis of quantum theory, viz. the discreteness of the level scheme.") Indeed, quantum rules allow for a great diversity of stable structures only a few of which can be envisioned realizable at zero absolute temperature in a classical world (a plasma, say). Equations of motion in the quantum mechanical description of nature are mute. Unlike their classical counterparts which allow a visualization of the continuum of activity, in the quantum cases we have to visualize, instead, a set of trapping and releasing operations as constituting the fundamental processes in the physical world. The anatomy of the Hamiltonian from which the equations of motion are derived, expresses without revealing it, an *underworld* of vigorous but pointless activity of virtualities. One can speculate that these virtualities in the vacuum fluctuations are essential triggers for the actualization of events and things. This is what is implied when the energy of the system is expressed in terms of the strongly anthropomorphic metaphor of *creation* and *annihilation* operators — the uncontrollable fluctuations between non-being and being. On principle any structure will have to be assembled on the basis of such sequential and memory-retaining operations. Yet little is known about initiating events, since a pure collection of the necessary ingredients will not assemble to form, say, a crystal even under the optimum conditions; there is a need for an initiating seed, a catalyst. I draw an analogy from the process needed to criticalize the fissionable material in a nuclear reactor or nuclear weapon with the use of a triggering radioactive material! The triggering process out of fluctuations may well be a rather general rule. Given a seed from a previous crystal, the appropriate ambient conditions and the supply of materials, one can conceivably expect that the "unitary collision matrix" of

Wigner[34] could, on principle, describe the "knitting" operation at the interface between initiated crystal formation and the colliding, and often trapped, particles which are added to the structure. And even for the most elementary of electronic devices (a junction, say), it would be quite involved to describe the process of diffusion needed to position the impurity gradients, the cutting of the wafers, and all these operations needed to establish a region of a non-equilibrium state for the electron flow.

It is beyond the present calculational means to develop a chronological, state shifting and augmenting algorithm which will include the "building" of a cell along with its subsequent interactions with energy. However, physical, macroscopic, evolutionary processes are developed by successive "trials." Each trial is influenced by the preceding one (this is the definition of a "Markov chain"[35]), so that all past history of the evolving process is represented at every succeeding step. G. Polya has described this as the *influence globale*.[36] Penrose (1970) in his *Foundations of Statistical Mechanics* interestingly introduces a set of five postulates on

---

[34] See his article "On the Probability of the Existence of a Self-Replicating Unit," in *The Logic of Personal Knowledge*, E. Shills (ed.), Glencoe Ill., Free Press 1961.

[35] Sequences of sums of independent random variables are the fundamental elements of stochastic processes. Two of the most important categories of stochastic processes are the *Markov processes* and the *martingales*; both are generalizations of assumptions about dependence relations among random variables. For instance, the first order Markov process implies a conditioning such that the value of the random variable X_t depends on its value X_{t-1} although independent of earlier times:

$$P(a \leq X_t \leq b | X_{t-1} = x_1, \ldots, X_{t-n} = x_n) = P(a \leq X_t \leq b | x_{t-1} = x_1).$$

Markov models have been employed in economics. A particularly relevant application is in the theory of growth of firms as in Steindl (1965); it may have a direct bearing in the biological realm and the physics of replication.

[36] Referred to in Steindl (1965: 153).

which he erects the theory of Statistical Mechanics. The fourth is the *Markovian Postulate* : "The successive observational states (sets of values for the indicators) of a macroscopic physical system constitute a Markov chain" (Ch. I.5). These observational states, however, are the results of trials which go through the stages of *preparation*, *development*, and *observation*; we recognize in this description much of what I attempt to articulate in the present thesis. The Markovian postulate is nothing but a formalization of the assembly of the capital-like structures which I have been proposing. Apparently then, it is within the process of coalescence and structure formation in which the Maxwell *demons* are active in the establishment of the Markovian behaviour of events. One may then conclude that what is finally studied and observed in thermodynamics and statistical mechanics is already frozen by earlier demons who built the earlier structures. Thus when the demons are invoked to sort out the particles in the style of Szilard and Brillouin (for which see section 5 later) it is already too late! The accidents leading to the "condensation" of matter and the interplay between the collective properties of structures and of single molecular events (collisions, energy releases, etc.) allow for the management of the energy processes required to address specific macroscopic requirements through the resulting capital-like structures.

This remark suggests that, had an algorithm been available, it might have been capable of providing the legitimation of thermodynamics in terms of the fundamental physical images that we have constructed of the world. In a reductionist approach, the various ansatzes, handwavings, and "superoperators" proposed are nothing but guesses and shorthands which for the moment substitute for a missing algorithm. Under these conditions, the importance of the *Second Law* of thermodynamics comes to prominence because of the accidents of nature which have been responsible for the formation of capital-like structures; to replicate, or quasi-replicate, these structures must successfully manage energy flows. This is the theme of the following chapter. In section 5.3 on measurement I will come back to the above ideas in order to underline the importance of structures and their historicity on the theory of measurement.

CHAPTER V

THERMODYNAMICS ($\Theta\Delta^{cs}$), WORK, AND CAPITAL

Ὦ δαιμόνιε, μὴ νουθέτει μ' ἀλλὰ πτέρα [1]

1. Work, all Work and Nothing but Work

WORK IS DEFINED AS Force × Distance = **F·dx**, the product of two primitive quantities. Work itself is not a primitive quantity; accordingly fundamental laws in physics do not allow for *work* anything but computational significance To measure work we often resort to *stopping power* techniques so that we may infer the amount of work from the condensation tracks or the heat generated in a macroscopic body during the passage of a moving particle[2] in a body or for that matter by stopping a large body (simply by applying

[1] Arist. 1436. Hey man! Don't give me no advice; wings will do.

[2] The Joule experiment is the macroscopic-classical version of the same idea. Often, when we inquire into the stopping action of a *macroscopic* system (a crystal, say) on a particle beam, then the work of deceleration gives a measure of the energetics of the event. Work transformed to heat gives information on the kinetic energy of the beam. Note that superluminal charged particles emit Cerenkov radiation in addition to exciting vibrational modes (heat).

the brakes!) The measure of work is significant when we wish to calculate what it takes to remove a particle from a solid — as during electron emission — or to add a particle in building up a solid, as well as to move a mass in a gravitational field. Work can be measured in different settings: a macroscopic piston moving in one (net) direction, a multitude of small "pistons" working in unison (coherently) — as in muscle tissue —, the massive motion of planetary atmospheres, in the *bricolage atomique* — the piecemeal build-up of a structure by adding brick on brick — as in the making of cells. Work assumes a special importance at all levels of the economic-biological world. Its acquisition or its loss are of great accounting and substative importance. These characteristics of *work* suggest that

> The Second Law of Thermodynamics formalizes the dogma that the premium in macroscopic energy transactions is *work.*

This section is written to support this contention. I have not determined whether this formulation is at all novel[3] but my intent is to underscore the issue of *scarcity* in *work.* I use the word "premium" advisedly because once work is available it can be fully converted to useless heat; it also implies a choice, although no deeper reason for this is intimated. Thermodynamics is thus endowed with a formal, utilitarian

---

[3] Although Berry *et al.* (1978) may have attempted a similar formulation when they state that "work sets the opportunity cost of fuel." But fuel on its own cannot develop work without the capital equipment which impresses constraints in the degrees of freedom. Work, in my view, cannot be obtained without the prior coherent (or even incoherent) assembly of capital. As a counterexample to Berry's statement we recall that the opportunity cost of uranium or coal is not set by the work it can produce but rather by the cost of capital needed to deliver this work.

attribute which is absent from the thermodynamics theories and textbooks; there is an occasional exception as in the case of Planck (1945 and IV.1 here). At the same time, I use the word *dogma* to suggest that the Second Law of Thermodynamics does not in itself provide a *justification* of the importance of work. Energy is a protophysical (non-definable) concept as is force and distance. By contrast work, a defined and computable quantity, is a measure of "majority-directedness." Thus, indirectly, thermodynamics places energy in the form of work on a pedestal and inherently claims that work, not energy, *is* the *measure of all things.*

Recall that the original formulation of the Second Law refers to an engine and the impossibility[4] thereof to supply work as its only output if driven by heat. It is, then, axiomatically assumed that work is an important utilitarian attribute and objective. Yet, there is no indication that work is a measure of goodness in the physical world or an essential measure for the formulation of the foundations of physics. The Second Law is heuristically comprehensible through the operation of an engine (capital) — the construction of which took another engine to accompish. An engine is not part of the foundations of physics, however. Further, a thermodynamic engine loses all meaning in microscopic physics although work can be a derived quantity at the microscopic level for purposes of calculation. On the other hand, the specifications *adiabatic, adiathermal, etc.* are only stenographic statements which describe controls over processes (diffusion, phonons, etc.) at the microscopic

[4] The injunction "impossible" does not imply a strict constraint (as for example in the statement: momentum cannot be exchanged in any away except through exact conservation) but rather a loose idea that directly depends on the particular environmental constraints of some collective.

level but which become concepts in the axiomatic formulations. Being a macroscopic discipline, thermodynamics does not (as it should not) deal with the underlying "perpetual" motion and/or vibrations of particles. It deals, however, with the impossibility of perpetual motion *machines* which cannot be energized out of the perpetual motion of particles.[5] Yet this is the essence of thermodynamics whose fundamental objective is utilitarian: the metamorphosis of heat, a random phenomenon, into work, a coherent movement in the large. (More technically it is the reduction of many degrees of freedom to very few). Work, unlike energy, implies the forcing of coherency in the motion of a collectivity of particles. Most of the particles must be moving in a preferential direction. With the premium placed in this unidirectional motion of the collective, thermodynamics was born.

In developing a microscopic origin for the *measure* of entropy (not the concept, as such, which has no microscopic meaning, not being an "observable" in quantum mechanics) in earlier sections, I was referring to the difference between two ways of taking an average: the average of a variation minus the variation of an average which is proportional to what is termed *entropy change* $\delta S = T\delta Q$; that is, the quantity which in the process of deriving some ophelimity [work], is given out as heat. Thus we may restate the earlier proposition by saying that *work foregone in energy transactions is a waste.*

---

[5] In this section I oscillate between the microscopic and the phenomenological macroscopic descriptions. The reader should note, however, that if the heuristics may be thus enhanced, the procedure may also be dangerous. It is formally entirely illegitimate to allow for any microscopic mechanisms in thermodynamics.

If then my contention about the utility and the importance of work is correct, I must look into the technological realizations which permit the supply of work with the least possible waste. The Carnot engine and the routines to which it is subjected is a paradigmatical case of such a realization. But in the more theoretical explorations in thermodynamics and statistical mechanics, these realizations are mentioned only in passing or opportunistically to surmount some logical obstacle in the exposition. In Schrödinger's book (1952) the procedure is transparent. After some abstract reasoning, he invokes "screws, pistons and what not" to effect the variations in the system — a thinly veiled macroscopic game. But it is these handles as well as the "heat baths" which are never discussed in detail, which should be included in the theoretical constructs in a comprehensive (and comprehensible) description of the thermodynamic-statistical mechanical process.

Although understandable because of its inherent beauty, it is still disturbing to observe the textbook disembodiment of thermodynamics. I do not propose to correct this defect in the present essay but rather to sketch a way out by underscoring the necessity of structures (which I call or associate with capital) interacting with energy for the development of work. These structures may result out of accident[6] or out of purposeful design, as in the case of the internal combustion engines. In all cases work output, the measure of thermo-

[6] *Ex post,* one can recognize that accidents of matter accumulation and periodicities can lead to processes which may be described much like thermodynamic engines. The earth's atmosphere is an example in which energy from the sun is received and reradiated cyclically while phase changes in the water occur and temperature gradients produced.

dynamic Utility (U_{th}), if you will, requires structures. Without these spatial "capital structures" thermodynamics would not be. They enable the interplay of energy events between the microscopic and macroscopic level and thus justify the emergence of the 2nd Law. Although never discussed in textbooks, capital construction is a resource consuming operation which, to achieve its design objectives [how insulating (adiabatic), heat conducting, heat resistant, thermalizing, separating (enriching[7]), cell feeding, breeding, etc.), competes with the energy needed for its own operation and repair. In particular, the 2nd Law would have no meaning had capital not been the structural element necessary for the delimitation of boundaries across which energy and materials transactions occur. In view, then, of the utilitarian requirements for work and nothing but work, it is essential to study the various capital structures for the extraction of the maximum possible work and the conditions under which this can be accomplished.

Work and *non-work* do not in themselves determine the scarcity of resources but only in conjunction with the capital equipment required to develop this work. Since, as I noted earlier, capital itself is assembled in time and piecemeal from work (the work necessary to add on the particles to make up the capital) there is negotiation with the future, a process which necessitates current abstention from useful *work* to assemble the capital now and obtain work in the

---

[7] Governments (and predators) have traditionally looked with consumate interest into anything that can be concentrated (whether it be uranium, alcohol, dried fruit, olive oil, grain, temple offerings) to tax, constrain or loot. It is simpler and cheaper to tax (or look for) the final rather than the intermediate product. For the utility of uranium enrichment see Dirac's 1941 computation cited in K. Cohen, *Theory of Isotope Separation as Applied to Large Scale Production of* U^{235}, McGraw, 1951.

future. In the macroscopic machines the correlation of the design blueprints and their intertemporal function is not immediately obvious; yet, as machines beget machines, their thermodynamic performance cannot be separated from their "genetics." In the microbiological level, by contrast, the intimacy of structure, function, and energy transfer mechanisms which is the basis of precision reproduction and work performance is recognizable. These remarks support the conclusion that the 2nd Law is not a fundamental law of nature at least in the context that we understand these laws. But that it is a result of the historicity of the K-E interactions described in the chapter on economics. Indeed, as these interactions result into the making of work producing machines, the constraints of the 2nd Law become apparent when the K-E systems are subjected to routines or programs. The nature of these routines must be sought in heuristic thermodynamics and its history.

Carnot's great contribution was the idealization of the processes involved in the supply of work. If two *finite* bodies with respective temperatures T_2 and T_1 ($T_2 \geq T_1$), are brought into contact, we observe heat transfer from the warmer to thc cooler body and an equalization of their temperatures; experience shows that the *status quo ante* (SQA) cannot be recovered on its own; the phenomenon is *irreversible*. Idealizing the situation, just a bit, we may conceive of the two infinite bodies, each with infinite heat capacity to have been brought into contact, while a minute, infinitessimal amount of heat ($\delta Q = \delta E = T\delta S$) is allowed to flow from one to the other. The entropy of the whole system increases by the amount $\delta E(1/T_1 - 1/T_2)$ where δE is the energy transferred.

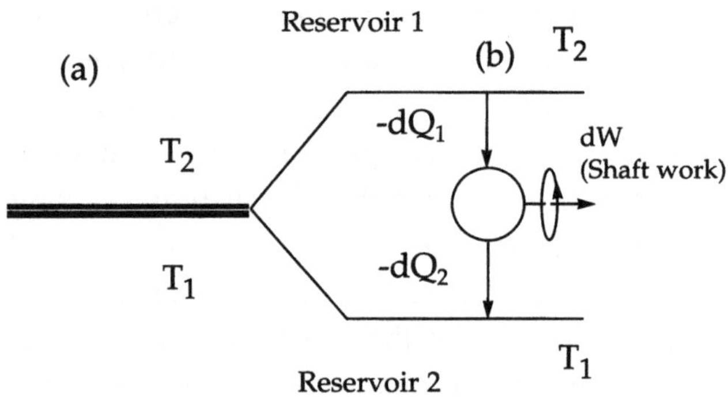

Figure V-1. (a) Heat Transfer at an interface and (b) the Carnot Engine with a shaft for work.

In the above structure (with reservoirs in contact) work cannot be developed. Thus, Carnot kept the two bodies physically separated by a space which permits certain manipulations: one can imagine a tube filled with a gas and a piston which is first brought into contact and equilibrium with the warmer medium of temperature T_2 so that it may absorb an infinitessimal amount of heat; this energy transfer leads to an expansion of the gas inside the tube (the *working fluid*, as it is known) while the temperature remains constant. Then the system is left to expand adiathermally (adiabatically) i.e. with $dQ = 0$. Brought into contact with the lower temperature reservoir, the gas is then allowed to transfer energy across the boundaries isothermally: this causes a contraction of the gas which is followed by a further adiabatic compression. The working fluid — with the piston following the exercise — reaches the

SQA except for something quite important. There is now a net amount of work developed. Energy conservation requires that the amount of $-\delta Q_1$ taken from the hot reservoir is equal to the amount of work dW delivered to the environment plus the heat δQ_2 rejected to the cold reservoir. Thus $-\delta Q_1 = \delta Q_2 + \delta W$. In the accounting, outputs from a system such as the reservoirs are taken as negative while inputs (to a system) are taken as positive. For the reservoirs, being equilibrium systems, energy conservation gives $\delta Q_1 = \delta E_1$, $\delta Q_2 = \delta E_2$ where δE_1, and δE_2 are the changes in internal energies of the hot and the cold reservoirs respectively. Further $\delta Q_i = \delta E_i = \delta S_i$, so that $\delta W = -\delta Q_1 - \delta Q_2 = -T_1\delta S_1 - T_2\delta S_2 = -(T_1 - T_2)\delta S_1 - T\delta S$. where $\delta S = \delta S_1 + \delta S_2$, the total entropy change in the system. Thus in terms of the heat extracted from the hot reservoir we can, finally, write

$$\delta W = \underbrace{(T_1 - T_2)\frac{-\delta Q_1}{T_1}}_{ideal} - \underbrace{T_2 \delta S_2}_{real}.$$

The maximum work is delivered when $\delta S' = 0$, which corresponds to reversible changes at zero entropy change for the reservoirs, the working fluid (which undergoes a cycle) and the environment, in other words, the zero entropy change of the overall system enclosing all the above elements. These reversible conditions never exist in reality althought they can be approximated sometimes very closely but at an increasing (economic) cost. The last term may, therefore, be considered as the term which fits ideal conditions to reality. This complicates matters enormously and leads to corrections to include the inherent propensities of real systems. The theoretical idealizations must be brought into contact with the realities of the engineering world. For example, to design a system that will transfer net energy isothermally or adiabatically one must respectively select appropriate exchange surfaces of highly heat conducting or perfectly insulating materials. These requirements entail the use of resources all the more expensive at the margin and scarce as the ideal is to be approached more closely. To decrease the entropic degradation, tradeoffs with capital are thus needed. Physical limitations in the peformance of materials, the quantities of materials needed, and their cost are the constraints as indicated in Appendix 2 on thermoeconomics. The temperatures cannot be increased indefinitely, neither the speed of heat transfer; copper is scarce.

I now return to the questions of irreversibility. The literature on the subject is quite extensive; however, it deals mostly with the *phenomenology* of irreversibility and as such is only tangentially of interest to the present essay. Here, we are interested in the "first principles" which (we think) must be responsible for the transformation of the fully time reversible equations of motion — the foundations for the description of the physical world — to sets of equations which constitute the foundations of thermodynamics. *All* the "fundamental" microscopic equations of motion in physics as well as some macroscopic ones are insensitive to the direction of time; they are "time-reversal invariant" (see IV, ftn. 27). Applied to theoretical descriptions they will carry events forward or backward in time with equal ease. Retrodiction is then possible as is prediction — at least on paper. As shown in Chapter II economic equations have similar characteristics. This property is referred to as reversibility and implies the ability of mechanical paths to be retraced. It is, of course one thing to *think* of time-reversal and of retrodiction and quite another to actually bring it about. The problem can become quite intricate if the processes of interest in addition contain energy transformations at the atomic and the molecular level. The general question then becomes whether we can attain the status quo ante (SQA) and at what cost. To reach the initial state would, in practice, require the design and construction of special purpose equipment and various amounts of work inputs. Thermodynamic experience as abstracted in the laws of thermodynamics, implies that to reach the SQA would need more *and* never less *work* than the system gave out while changing state. This effect is referred to as irreversible. It is really a question of *reproducibility* of some macroscopic state, *not* reversibility by going through the same exact steps backward, although that possibility might not be excluded. An example might illustrate the discussion.

The proverbial egg beaten to make an omelette can become an egg once more, not by reversing the movie camera which recorded the sequence of events, but by feeding the ingredients into an *egg generator* machine ... a chicken. The difference between the original and the reproduced egg might be extremely difficult, if not impossible, to discern and measure (atom for atom), while the probability might be finite that after a limited number of trials, eggs, thermodynamically and phenomenologically equivalent to the original, might appear.[8] The reconstitution of the eggs movie-

[8] Apparently the idea is not that original as I thought. I was alerted to this by the pangnost Daniel C. Dennett who in his *Darwin's Dangerous Idea*, Simon & Schuster 1995 referred to the "unforgettable example," of unscrambling scrambled eggs in R. L. Gregory, *Mind in Science: A History of Explanations in Psychology and Physics*, Cambridge, Cambridge University Press, 1981. I thank Prof. Gregory who believes it was an original idea of his (letter of 30 October 1996).

like, on the other hand, has an effectively zero probability of occurance, whatever the resources used for the procedure would require the identification and relocation of the original particles in the egg, a practical impossibility without … demonian help. The problem centers on the "egg generator" idea, i.e. the transfer of the question of irreversibility and retraceability of paths in a many body system to that of the physics of the formation of capital-like structures which now, as with the egg generator, have the function of production and reproduction. The idea of *coarse graining* in statistical mechanics might, in this case, refer to the slightly different eggs that would be produced by the "egg generator." The ensemble of these slightly different eggs would then be responsible for the *variational* process of biological evolution. Penrose[9] approaches a related problem when he refers to a broken watch and its reassembly: "the reason," he writes, "that I had a watch earlier is not that these precise correlations exist in the *future* boundary conditions, but that something in the *past* (say a watch factory) had a lower entropy than it might otherwise have had. Likewise, the reason for the precise correlations in the particle motions of the shattered watch can be traced back to the factory, not the other way around." *Thermodynamic* irreversibility has no meaning in the microworld of particles because it refers to processes which are compared against the possibilities of work itself — a macroscopic utilitarian measure. But note that one may describe a quantum-mechanical transition as

[9] Roger Penrose in "Singularities and time-asymmetry," in *General Relativity*, S.W. Hawking and W. Israel, eds. Cambridge University press, 1979: 787. Penrose makes the intriguing proposal that one could impose on the initial singularity of the universe a time-asymmetric initial condition — namely, the vanishing of the Weyl tensor and that this is related to the second law of thermodynamics via the field's contribution to the total entropy density of the universe.

irreversible in the sense that, left on its own, the products of the transition will not recombine to obtain an SQA — unless, of course, we build "recombination machines" of macroscopic proportions. This requires set-ups of considerable ingenuity and specificity. The chemical processes do this in the bulk and they entail heavy work penalties; the aforementioned egg generating machine is no exception.

In conclusion. Maxwell's objcctions appear to remain valid; the *ad hoc* introduction of "dissipative" components, right at the microscopic level may lead to desired results but is not justified on first principles; further, it is operationally sterile as it leads to no new insights or problem solutions. My thesis then is that the 2nd Law of Thermodynamics, as is presently articulated, *may also be* a fundamental law of nature but if it is not, attempts to derive it from first principles leave much to be desired. Indeed the discipline of thermodynamics, perfectly logical within its own confines, needs an extensive program of review and legitimation. From the discussion this far, it appears that the basis for such a program will require a) an elaboration of the capital-energy interaction and the origin of capital and b), an elaboration of real engines[10] that to work require a non-equilibrium environment. Thus a reductionist program in thermodynamics can only be devised by accepting the dialectics of value in *work.*

[10] There may exist a precedent noted by Truesdell *et al* (1978) who argued that F. Reich introduced axiomatic thermodynamics based on Carnot cycles in 1856 so that "the theory of heat engines is not an application of general principles but the source of them."

δῶς μοί πᾶ στῶ καὶ τὰν γᾶν κινήσω [11]

2. *Biological Work, microscopic engines*

This section should be read in conjuction with the chapter on biology. There, my interest focused on the origins of cells and the roundaboutedness of the reproductive process. It is now appropriate to recall the K-M biological model and view the cell as an engine that develops work for motility and a processor of materials. While it is believed that the study of the mechanisms of the molecular motors will shed some light on the origins of the cells, the justification for this section is to further support the view of the biological engines as capital devices. My intention is not to dwell on the subject with professional ambitions but, rather, within my own very limited knowledge of the subject, to identify the key elements that appear to be of generic value in support of this essay.

While this is written, the physics and chemistry of muscle or muscle-like cells is undergoing intensive research, supported by highly sophisticated technological experimental[12] work on motor proteins and their molecular mechanics; these promise the resolution of many of the mysteries of muscle function although much inference is still the way to form judgments. No one seems to have watched an *in vivo* molecular motor operate and no definitive theories are

---

[11] Give me where to stand and I shall move the earth. Archimides.

[12] For a description of recent remarkable experiments see, for instance, J.T. Finer *et al* "Single myosin molecule mechanics: piconewton forces and nanometre steps," *Nature*, **386**, 1994: 113-119.

presently available. Indeed, the exact mechanisms for muscular contraction and triggering, as well as the necessary energy transactions, are still speculative as can be deduced on perusal of the literature.[13]

This far we have addressed the thermodynamics of engines that have macroscopic dimensions. When asking, however, about the limits of real machines, as these get smaller and smaller, one has to turn to biological cells— constructions that have been tested in the evolutionary scramble for survival. Cells are chemical factories, of course, but the internal motility of their parts or motility of their bodies are key elements for function (including reproduction) and resource utilization. Muscle is the principal work-producing cell.

Recall that in macroscopic machines (engines) the development of work is accomplished

> a) directly through the pressure developed from an autocatalytic chemical reaction (explosion) that acts on a piston (or fan blades) — as in the case of an internal combustion engine (or turbine) during an isothermal expansion — and indirectly through a working fluid, as in a steam engine (turbine) or,

---

[13] See, for instance the article of Eisenberg and Hill (1985). For a recent report on these findings see J. A. Spudich, "How molecular motors work," *Nature*, **372**, 8 December 1994: 515-518 and current references therein. For a general but fast obsolescing review see for instancce *Encyclopedia of Science and Technology*, McGraw-Hill 1972. Lev Alexander Bl[i]umenfeld, *Probleme der Molekularen Biophysik*, Berlin, 1977 and, *Problems of Biological Physics*, Springer, 1981, E. Broda, *The Evolution of Bioenergetic Processes*, Pergamon, Oxford, 1975.

b) through an entropy machine that utilizes the system's free energy to raise the potential energy of particles that participate in energy-donating and accepting chemical processes (batteries, fuel cells, etc.); also electric motors.

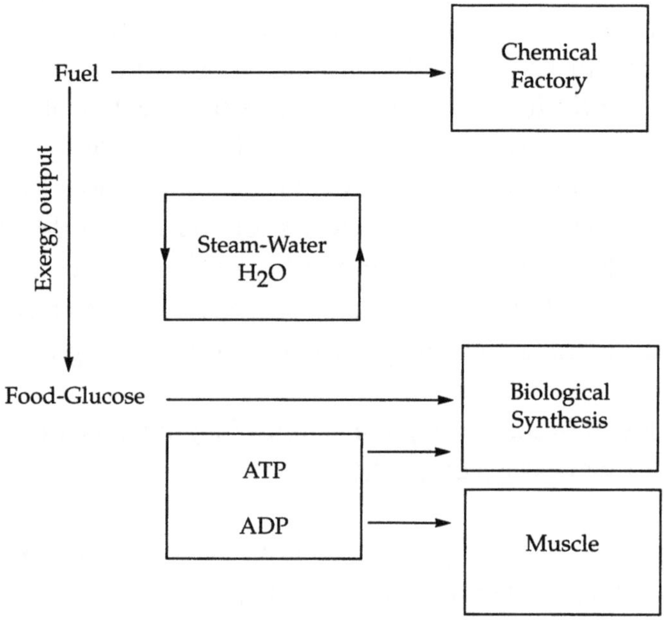

Figure V-2. The Steam-Water and the ATP-ADP Cycles (ftn. 14).

Neither of these mechanisms can, even approximately describe the performance of the muscles. Furthermore, in both cases the management of the flow of fluids and particles is done with the openings and closings of ports and valves at the expense of energy that would otherwise go to useful work. As machines get smaller and smaller, with a proportional decrease of the primary energy inputs, the energy lost for control functions (the overhead) becomes large by comparison with the useful work output; at some, limit fluctuations and the quantum uncertainty restrictions set in and make control altogether impossible, if at all meaningful.

What then is the connection of biological and mechanical engines? None in structural content, much in terms of "working fluids." For some useful comments on this see Grassmann[14] who compares the steam-water cycle with that of the ATP-ADP cycle. In the latter case we have two outcomes: biological synthesis (assembly) and chemical exergy, and (muscular) work. In the former case we have work output. As has been noted before Kováč,[15] an engine as well as a cell is composed of the working fluid part and the constraining system with very few degrees of freedom. Few points in the *p-q* phase space are accessible while the system is to be found in relatively few microstates. High barriers separate regions. The similarity ends here however, since work is derived from atomic collisions of an energetic gas in one case, while in the second through the slow and "rowing" reversible action of molecules that can be in one of two possible states. One should, perhaps, avoid trying to find *structural* parallels with macroscopic engines and instead return to thermodynamic basics. We have seen the approach of Gibbs in the development of the *grand canonical ensemble* and its application in the generalization of the idea of availability: it consists of dividing space into two volumes separated by a "membrane" that can serve various

---

[14] G. Grassmann, "Thermodynamik des Lebens aus der Blickwinkel der technischen Thermodynamik und die Exergie," *Naturwissenschaften,* 71, 1984: 335-341.

[15] L. Kováč, "Overview: Bioenergetics between Chemistry, Genetics, and Physics," *Current Trends in Bioenergetics,* **15,** 1987: 331-372. This overview incorporates in its text valuable comments and insights. In addition it provides a much needed synthesis which describes open problem areas.

boundary conditions for heat and materials transfer. This is also a model of a cell (see Appendix 1).

The metabolic system provides the energy to assemble the engine and incorporate in it the replicative mechanism. But the geometrical variability and the movement of the whole in its positioning of the "engine" permits the acquisition of material and energy inputs. Therefore, any study in origins of cells must consider the "machine elements" and their evolution with specifications that may have been the precursors of the muscles and nerve cells. These considerations do not seem to have interested the specialists as they could develop independently of replicative mechanisms. Functions then (and their mechanics) such as membrane permeance and membrane mechanical dynamics, become key elements for a larger synthesis; these biological "components" permit a structure to "reach out" whether induced by external forces (gradients) or by internal feedback.

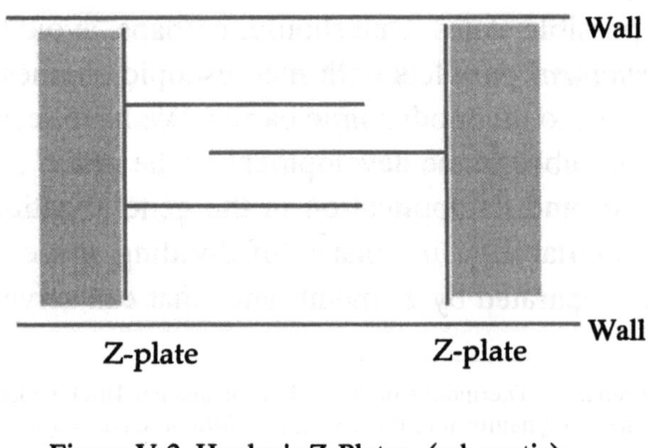

Figure V-3. Huxley's Z-Plates (schematic)

Muscle elements. The only accepted and experimentally confirmed mechanism is the one proposed by Huxley (1966, 1969) on the interpenetration of the myosin and actin filaments. The present models on the mechanics of muscle are founded on this pioneering work based on macroscopic muscle fiber experiments. Myosin, in the form of bipolar thick filaments, is the basic motor molecule that interpenetrates a thinner actin filament. That the various thermodynamic (macroscopic) theories and principles do not provide clues as to the micromechanics of muscle cells and their extraordinary efficiency, has been discussed by McClare (1971, 1972) and Gray (1975).

In muscle contraction, chemical energy deriving from the hydrolysis of ATP is transduced to the relative "sliding" of actin and myosin filaments. It is generally accepted that for every such event, myosin binds to actin; this causes a conformational change (knee jerk-like) and then detaches. The details of this mechanism are still obscure as the experimental data are derived from stripped molecules without their cellular environments and collected at high densities.[16] One can visualize the muscle cell in many different ways. In some cases it represents the configuration of a biological cell itself and in others it may represent a repetitive structure. But, in general, it seems that it is the deformation of the myosin macromolecule that leads to work in terms of a pull-resile cycle energized by the ADP-ATP transition and induced by the Ca^{2+} atom. As a consequence the muscular units (the cell) can be visualized as a sarcomere composed of thin and thick filaments, sliding with respect to each

[16] Recall that Bohr commented that the *in vivo* experimentation at the macromolecular environment is a contradiction of terms; Bohr, N., *Atomic Physics and Human Knowledge,* Wiley, New York, 1958: 20.

other through the function of knee-like molecules that push [propel, oar, paddle] the filaments; in effect a box defined by the Z-plates and having movable porous walls admitting and rejecting molecules, atoms, protons, electrons, etc.The myosin-actin system looks more like a microscopic arthropod within a tube. One must then suppose that in each "cell," ATP molecules deliver energy in the form of work while the molecule itself is converted to ADP and a free P_i atom. This hydrolysis must then be followed by regeneration of the ADP molecule back to ATP through the energy inputs into the cell and through the walls which reestablishes the conditions for the next cycle of work release.

With this model at hand, the efficiency of muscular operation can be accounted for, while the thermodynamics must be considered globally by including the *external* low efficiency energy circuits (the power supply) required for the recycling of ADP and the rejection of heat generated by the interfilamentary friction. Thus, to envision the ATP-ADP cycle all its externalities must be included, that is, all the supporting biological system incorporating muscle. In the particular case of the muscle, we could, then, envision the mechanism as consisting of an enrichment process which accumulates, at distributed sites, molecules which are energy rich (ATPs) and which, on command, dissociate, while the released energy used to change conformation of the myosin; the resulting configuration changes act like "micropiston" (cells walls; Z-plate) and the collective effect — properly modulated — results into the observed macroscopic work. As a result of these conjectures one can, then, propose that the *macroscopic* muscular activity results from a coherent summation of a large number of participating cells (sarcomeres) as they change state. The coherence of the collective contraction must, however, come from some centrally directed signal to each of the individual cells. How the signalling is accomplished in the muscle to obtain the requisite coherence remains an open question but without doubt the Ca^{2+} ions are involved.

On Demons once again. Right after the epochal work of Huxley, the question arose as to the specific physical mechanisms of muscle action. After fruitless explorations, McClare (1972) remarked that the observed processes have similarities with the operation of lasers and that "apparently, the ATP molecules are made or used up, one at a time." What, for instance, seems to have baffled Gray (1975) is an apparent mechanism with the attributes of a Maxwellian Demon operating at the cellular level; a Demon, nevertheless, immune to thermal fluctuations and without any evidence of a *microscopic* thermodynamic engine working on a cycle! The exercise could have remained academic had it not subsequently inspired an imaginative, albeit stylized, quantum-mechanical model of cellular contraction of Gray and Gonda (1977). These remarks led them and Blumenfeld (1978) to propose the adoption of a quantum-mechanical box-model of a "cell" (the muscular sarcomere unit) whose walls — being unconstrained — can experience a net force and a dislocation when the transition of ATP to the state ADP + P_i occurs. The attempt was misdirected as it centered on the identification of a work-producing system in the form of a miniature thermodynamic engine working on a cycle within the Huxlean filamentary model for the "mechanism for molecular contraction." But this inquiry opened other questions such as "what is the smallest Carnot engine that can be constructed to operate on a cycle" — a subject referred to below. All this prompted me to look into the Maxwellian demon at a time when considerable interest in the issue has been exhibited by the information-theoretical community and the Leff & Rex (1990) reprint collection. Immediately one focuses on the Szilard box in an attempt to see its possible connection with biological reality.

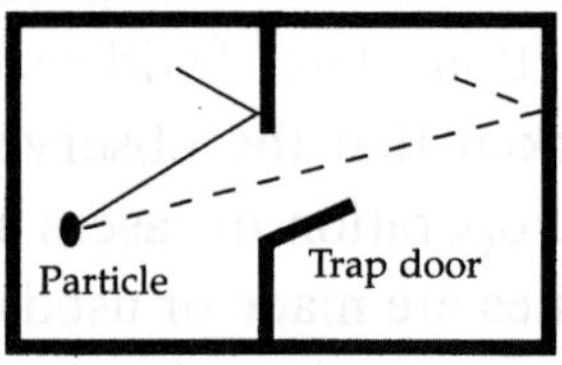

Figure V-4. The Szilard Box

The Szilard thought experiment and work: Szilard begins with the classic expression pV = RT of thermodynamics, an expression that is phenomenological, macroscopic, and relating to a continuum. Then, since work is given by p×dV (pressure×change in volume) he integrates from V/2 to V to obtain the celebrated expression kTln2 where k is the Boltzmann constant and T the temperature. Suddenly he changes outlook. His physics become microscopic and he refers to *one* particle inside the box which, on hitting a partition, moves it while the experimenter goes through the routine prescribed by thermodynamics (move slow, don't increase entropy, etc.). The Szilard box can have any dimensions ranging from infinitessimal to very large. Single particles are endowed with temperature, they interact with frictionless "pistons." As gedanken experiments go the structure studied for the operation of the Maxwellian Demon is one that can be thought to be as small as needed. Realistically engines cannot be indefinitely miniaturized, however. Biological structures in fact are alternative realizations of microscopic engines but then we must be prepared to look into them from a different perspective. We should not be looking for pistons, valves, fans, rods, fuel injectors etc. i.e. the components that are needed to assemble an engine. The biological engines differ from physical engines, as Kováč notes, in one key respect: they are protein machines i.e. cooperative structures. Unlike the physical machines in which a component can be replaced to restore operation, the biological machine being one unit has to be replaced by a new one. The discussion of the Maxwellian demon becomes meaningless. To save the experiment, the analysts (aware of the incosistencies) call on Gibbs' ghost, the ensemble i.e., identical copies of the one particle box. In between, the particle bounces indefinitely around in a perpetuum billiards machine. No wall interactions, no interparticle interactions. Just a minute interaction among the copies of the box! The gendanken experiments are, of course, confined to physics where unreality can be a teaching aid. In macromolecular biology, where working systems are encouraged, the ghost of the ensemble has no place. The Szilard box and the trap doors substitute for the control system of an actual engine (the valves, rods, gauges, etc) with thought; it is a thought experiment and its consequences are thoughts too. Gedanken-machine elements operate as bona fide real machines only rhetorically or stylistically.[17] The obvious question is to see how the

[17] It will be noticed that I avoid the rhetoric of information as entropy or vice versa. For some thoughts on the subject I have provided Appendix 4. The reader should, however, consult the collection of papers on the subject edited by Leff and Rex (1990). Operationally the association or equivalence of the two

Szilard box might be adapted to biological needs. Zurek[18] provides a quantum analysis of the Szilard engine. He follows the prescriptions of the quantum exercises of gedanken boxes with appropriate boundary conditions for the wavefunction ψ and assumes the cylinder to occupy the space between -L/2 and L/2. The textbook calculations follow easily but then Zurek implies that all virtualities in the box will be weighed by $\exp(-\beta E_n)$; why virtualities are so weighed is a mystery unless, on the "collapse of the wavefunction," they interact with the reservoir next door.[19] Following Zurek we might have to insert, instead of a partition with its quantum potential well box, a macromolecule that can change configuration pushing or pulling the walls and thus developing work:[20] indeed all current findings indirectly support the view that work at the micromolecular level is produced through the conformational changes of mole-

---

does not affect the making of science; but does it even lead to inventive thinking? K. Denbigh, "How subjective is entropy?" *Chem. Brit.* 17, 1981: 168-185. Critical of the equivalence he agrees with Hobson to the effect that one should emphasize the "entropy of the data" not that of a physical system. But one wonders what, if any, contribution to science has the remastication of the Maxwell demon provided.

[18] W. H. Zurek, "Maxwell's demon, Szilard's engine and quantum measurements," *Frontiers of Nonequilibrium Statistical Physics,* in G. T. More and M. O. Scully, eds. Plenum Press 1984: 151-161; also in Leff & Rex (1990). Nevertheless all these belong to the family of homework "exercises" in solving the Schrödinger equation under various theoretical, idealized, unrealizable constructions that imitate the macroscopic images. Stylized (as the economists would say).

[19] For the meaning of "virtualities" and "collapse of the wave function" see the next section on measurement.

[20] Whatever the model, however, work is developed at the microscopic level along the lines of a mechanism that is related to the one observed in the electron-nucleus interactions in molecules and calculated in the Born-Oppenheimer approximation; or, one discussed in the properties of an electron in a potential well with movable barriers. For an interesting discussion on the physics of forces of attraction between adjacent solid bodies in terms of electromagnetic fluctuations in the intervening space, see L. D. Landau and E. M. Lifshitz, *Electrodynamics of Continuous Media*, Pergamon Press 1960: 368-376.

cules. Nowhere does one see a theoretical quantum mechanical description, however. The driving force $F = -\partial E/\partial X$ is related to the energy level E. Large molecules may present dynamics that are difficult to treat quantum mechanically, especially when they have alternative conformations (as in tthe case of the ammonia molecule which changes states between two equilibrium configurations during transitions while radiating in the microwave region). The far more complex haemoglobin molecule has two states that resemble the clicking of the mexican beans. And more extreme is the myosin-actin complex with its large scale motor action at the nanometer level. All must fall in the same category of forbidding problems where theoretical approaches cannot yet handle conformational changes. The myosin-actin machine is a two state device; every time a drop of Ca^{2+} is pippetted in the quantum box work is extracted. To do it again the molecule has to be relaxed, at which time entropy increases.

To summarize: Biological systems utilize machine elements unlike any macrophysical machine that we know. In muscle the mechanisms do not depend on brute force to develop work. And to reduce entropy, biological systems develop folding surfaces of enormous areas into small overall dimensions (small volume to surface ratio). Judging by availability considerations where to reduce entropy production increased interaction surfaces are necessary, Grassmann points out that natural selection has found an economic way to develop intestinal walls that because of their folding and microstructure provide an effective $200 m^2$ of surface.[21] Further, "the eukariotic genome reappears as a busy workshop, overfilled with raw materials, spare parts, broken products, and junk of all kinds, and furnishing a variety of finished products."[22] Among these, one apparently encounters a plethora of molecular motors, such as the cytoplasmic "dynein."[23] Myosin is thus not only related to muscle but to the architecture and the cytokinesis of the cells themselves. The apparent ubiquitousness of the myosin, kinesin and other motor molecules, within cells and in muscle, allows one to surmise that, in the early phases of cell construction these molecules were the key elements, first in structuring the boxes (cells) in which reproduction occurred, and then allowing for their motility to look for free energy and accomplish reproduction. We may conjecture, then, that we have in our hands machine elements in their ultimate miniaturization, i.e. the structural units of natural capital. In general the reproducing cell is the basic biological structure that plays the role of capital.

The next section is on Measurement. This far I have argued for the necessity of adding structural and economic considerations to thermodynamics without, however, touching the actualization of this process. The question is how to connect the theoretical outlook with the actual construction and experimentation. This I set out to do by noting some of the yet unresolved fundamenatl problems in physics and cognition while altogether skirting the philosophical issue of the theory-experiment nexus. To provide a setting I also introduce a set of categories of experiments in Appendix 5.

---

[21] For the discussion of *availability* see Appendix 3. For Grassman, see ftn.14 of this chapter.
[22] Kováč, *op. cit.* p.347.
[23] That Spudich estimates to be close to 100 in number! (see ftn. 13)

3. Measurement in Physics

Unlike geometry and arithmetic, where the foundational axiomatics are essential for their development, physics is surprisingly (and, perhaps, all to the good) nonchalant about its foundations. Characteristically, physicists, whether theoretical or experimental, work with an interplay of techniques, guidelines and heuristics — symmetries, renormalization, conservation principles, gauges, group theoretical relations — strong imageries, intuition, not to mention "magic."[24] With few important exceptions,[25] physics works incrementally — by exploration and improvability — and at the margin, and is principally concerned with the making of things (inducing events and happenings — possibly in the form of "thought experiments") rather than with the search of ideals — although there is important and productive work going on in this direction since the time of Plato.

An inventive experimentalist has his own kind of visualizations (anschaulichkeiten) and models of interaction which assist him in planning, executing and interpreting an experiment. Experimentalists are continuously on the lookout for developments in technology (electronic detection, amplifi-

---

[24] As Penrose (1989) correctly characterizes the theoretical machinery in quantum mechanics.

[25] This is not the place to discuss the origin of ideas and especially those with revolutionary content: Maxwell's unification of electromagnetism, Einstein's introduction of simultaneity and invariance, Schrödinger and Heisenberg's quantum formalisms, the Darwin postulate on evolution and selection, are examples, so that I prefer to refer to the "unique exceptions." I use the expression "at the margin" intentionally to imply the economic notion that as more is demanded of a good (here, research within a paradigm) more expenditure of effort may be needed for fewer and fewer important results.

cation, radiation sources, filtering, computation) which might allow them to develop and then apply alternative or more sensitive, procedures in the quest for the validity of physical laws down to more and more elementary levels. Others imagine new interactions, conjure new possibilities, seek new effects, and organize their setups in search for novelty and the realization of "virtualities," at least in the spirit of the Copenhagen School and its offshoots. These functions are usually covered by the word "measurement". Although the literature abounds with experimental descriptions and results, a universally accepted theory of measurement is missing. Indeed, the theory of measurement is the fundamental epistemological problem of modern physics and is an area which is still evolving as if though little was needed of it to make advances in physics![26]

As in many other subjects in the humanities, physics is experiencing a strong wave of interest in its own history; this is, perhaps, symptomatic of the introspection and uncertainty of "postmodernism": the critique of the old dogmas exposes their weaknesses. By contrast with the humanities, the deconstruction, as it were, has constructive objectives in an attempt to probe deeper and resolve outstanding puzzles. The renewed interest in these themes derives not only from basic philosophical endeavours but also because physicists have come to realize that successful though a theory may be in predicting events and providing means of computation, it leaves too many questions open especially in the

[26] Again the literature is immense and the reader is once more referred to Wheeler and Zurek (1983), a recent collection of key contributions to the field which contains a guide to further reading along with an extended bibliography. There is also extensive literature in the history and philosophy of science on the subject of measurement which the reader may want to consult but which is not referred to here.

epistemology of measurement, in cosmology, and in high energy physics; more recently scientists are also getting interested in the operation of the mind and consciousness. It is in this spirit that one can profitably study recent work.[27] Indeed, how can one articulate the problematique of measurement without, at least, an awareness of these issues and questions? Before going on with my objectives I will make a short comment on how I position myself in this respect. For my own objectives and quite tentatively, I accept the view of the "multiple drafts" proposed by Dennett and implicit in Edelman's[28] process of "reentrant signalling" as a mechanism leading to consciousness but consider it incomplete, as it does not obviously steer to the virtual cognitive center that I referred to in the introduction. To cover this gap I conjure of a process very much like the one used in what is known in quantum electrodynamics as *renormalization*: it, in fact, involves a summation of an infinite series of events that formally diverges unless doctored to remain finite. But once doctored, it provides superb agreement between theory and experiment.[29] In the brain we may

---

[27] Without exhausting the list see: M. Lockwood, *Mind, Brain & the Quantum — The Compound 'I'*, Oxford and Cambridge MA, Blackwell 1989; R. Penrose, *The Emperor's New Mind — Concerning Computation, Minds, and the Laws of Physics*, and *Shadows of the Mind — A Search for the Missing Science of Consciousness*, Oxford University Press 1989 and 1944 respectively; H. P. Stapp, *Mind, Matter, and Quantum Mechanics*, Springer, New York, 1993. The above material is controversial because exploratory; however, it is highly recommended for superb tutorials on various subjects physics textbooks pass by with handwavings.

[28] Gerald M. Edelman, *The Remembered Present — A Biological Theory of Consciousness*, New York, Basic Books, 1989, and Daniel C. Dennett, *Consciousness Explained*, Little Brown & Co. 1991.

[29] The field of renormalization is by now huge since the property of renormalizability is considered to be the fundamental constraint for a physical theory to be effective. See F.J Dyson, "S Matrix in Quantum Electrodynamics," *Phys. Rev.*, 75, 1949: 1736 in particular section IX, "Discussion of Further Outlook." Also instructive is Tian Yu Cao and Silvan S. Schweber, "The Conceptual Foun-

yet conceive of an Elelmanian process of reentrant signalling that continuously refreshes memory. The sum of all these may, in fact, be what we call consciouness. After all, renormalization is a product of the mind, perhaps even mirroring its internal techniques.[30] The belief is, however, that consciousness is not due to just the additive behaviour of neurons. Some guess that it is due to "cooperative states among networks of neurons." Others that association of two features necessitates to bring the two pieces of information to the "same locus," etc. One may conjecture that consiousness is a physical state altogether different from those currently described that suddenly snaps in the volume of the brain. Similar conditions of getting into a different state have been described in physics. The state of animated consciouness, I conjure, is this "resonating," "ringing," "tensed" alertness; a state that has many common characteristics with a phase transition in critical fluctuations or of collective effects as in superconductivity, a sponaneous symmetry breaking, whereupon something incoherent acquires structure in physical or phase space. Could it be that this is what is perceived as the virtual cognitive center? In any case here we are concerned with one aspect only, namely, the epistemology of measurement and, in particular, that part that relates to the foundations. In addition to

---

dations and the Philosophical Aspects of Renormalization Theory," *Synthèse*, **97**, 1993: 33-108.

[30] Related conjectures have been proposed in the past. For instance C.I.J. Stuart, *et al.*, in "Mixed-System Brain Dynamics: Neural Memory as a Macoscopic Ordered State," *Foundation of Physics*, **9**, 3-4, 1979: 301-327 produce a "mixed-system" hypothesis in which some functions of the mind are strictly quantum mechanical in the large, while I. N. Marshall, in "Consciousness and the Bose-Einstein Condensates," *New Ideas in Psychol.* **7**, 1. 1989: 73-83, argues for a direct line from the quantum to consciousness not intermediated by the body. These are models worth having in mind as they enrich the alternative from which a synthesis will eventually result.

the above critique of experiment, from the point of view of physics and philosophy, there is another level of inquiry which I note, for purposes of completeness, although not of relevance, to the work at hand: it derives from the historical perspective and the epistemology of measurement[31] at the classical-logical level of discourse. At this level the critique of the foundations is absent but the methodology for measurement is taken to still be applicable.

Aporiae. But where do the puzzles in the foundations appear? Heisenberg (1958) gave an account of what he thought quantum mechanics meant in terms of words rather than, operationally, in terms of equations. This is now referred to as the Heisenberg "ontology." Heisenberg's operators, functions of spacetime, describe the state of the universe — the cosmos. The "expectation values" of these operators calculate definite values still applicable to the whole spacetime — the cosmos. Heisenberg describes these results as "objective tendencies" that anticipate the "actual events" — the apparitions, if you will. The operators describe waves that are then actualized as particle-like happenings. The Copenhagen school addresses these happenings as "collapses of the wave function." Indeed, by following the alternative (and equivalent) Schrödinger description of the cosmos through the ψ-function — a wave like metaphor that works *as if though* one had to consider vibrating strings — the "collapse of ψ" is one way of ... visualizing the transition to another state and observing an event.

[31] The practitioners are many but I only refer to an American Philosophical Association symposium (December 28 1988), "The Philosophical Significance of Experimentation," *J. Phil.,* XXXV, Oct. 1988: 507-527 with useful review articles: I. Hacking, "On the Stability of the Laboratory Sciences," 507-514; P. A. Heelan, "Experiment & Theory. Constitution and Reality," 515-524; P. Galison, "Philosophy in the Laboratory," 525-527.

Philosophically sensitive Heisenberg (1958: 147, 180) described this as paralleling the Aristotelian concept of δυνάμει (= potentia). In this model, the "happening" or the "event" is a discontinuous change in a Heisenberg state and this is only in preparation for the next event; i.e. a concatenation of events. But each event relates to a macroscopic domain and to globality — the universe partakes in the happening. This has important implications as it goes beyond the Copenhagen interpretation that remains within the confines of the Experiment-Observer Cartesian duality. Globality also poses other serious problems in instantaneity. And then, again, how and where does the whole chain of event-making begin? For "the where" there is the so called *Heisenberg Cut.* A cut must be made somewhere and thus "a quantum theory of the universe without an outside observer is a contradiction in itself."[32] Within the Cartesian model, measurement *is* the intersection of rational thought with the world outside it; as such, it is quite unlikely that a resolution of the fundamental problems is forthcoming. To make headway we follow the procedures of the protagonists (Bohr, Heisenberg, Dirac, von Neumann) and cleave the cosmos into Σ, the system under observation, and M, the observer (which includes instrumentation). The *Heisenberg Cut* is unavoidable although its boundaries may shift. Thus the system cannot be the cosmos itself. However, the next step is to bring in such slippery concepts as free will and consciousness. M, remains a "classical" entity comprehensible to "common sense." Common sense implies that the observer is a human being, a mind. But, then, how does one classify the Oklo nuclear reactor? Was it not a phenomenon since not (ordered and) observed by a human

[32] Haag (1992: 3).

mind but inferred by the archaeology of fission *ex post* ?[33] In this case the phenomenon becomes a macroscopically recorded event, part now of M.

In the same spirit, von Neumann (1932, 1955) opened a path, since followed, that makes a distinction between the Schrödinger equation and the collapse of the wavefunction. The former is a time-reversible deterministic equation of motion while the latter refers to a process that implies a "revelation" and realization of a virtuality. Physical systems change in time. Accordingly quantum systems change in two distinct ways: during a process of the *first* kind or one of the *second kind*. The second kind of change occurs in noninteracting (isolated) systems. The first kind describes change when a measurement and observation is performed on a quantum mechanical system. This process is the most controversial in the whole of physics but uniformely it is one on which the least is discussed in classes for, were it to become the focus of attention, not much else could be achieved; it could paralyze the learning and the making of science. Instead the von Neumann approach, known as the *projection postulate*, has been adopted without much resistance. The von Neumann doctrine refers to the *state vector reduction* or to the *collapse of the wave function.* To carry a measurement is nothing but the projection of the state of the system under investigation onto the eigenstate which corresponds to the resulting eigenvalues. The reduction is supposed to happen with perturbing and non-perturbing measurements since within quantum mechanics it is possible for a system to act without being acted

33 "The Fossil Reactors of Gabon," *Sc. Am.* **232**, 6, 1975: 41, with reference to a *Bulletin d'Information Scientifiques et Techniques*, French Atomic Energy Commission (no date).

upon. State vector reduction is a process that is supposed to transcend all ordinary physical interactions (such as involve only von Neumann's processes of the second kind). The idea is that measurement precipitates (in no time) a step change to an eigenstate associated with the "observed" state.

All quantum mechanical theoretical-mathematical operations follow this doctrine with no qualms and one suspects experimentalists adopt the model implied, however reluctantly. The experimentalist has to deal with apparati, with electronic amplifiers, signal generators and processors, reading instruments, traces, simple models of events; the von Neumann chain fits his objectives.

In a text noteworthy for its sensitivity to the lacunae in its foundations, Haag (1992) alerts the reader to the pitfalls to be encountered. However, in a separate article[34] he articulates what is outside the orthodoxy: can one isolate events with absolute precision? In following the von Neumann chain can one describe the universe as a set of events increasing in number as time goes on? This, he argues, would necessitate the introduction of irreversibility on a fundamental level[35] and the revision of the concepts of space and time. In quantum mechanics this is circumvented by limiting the class of events considered to the interaction of the "microscopic system" with a macroscopic measuring device. Then Haag remarks that once one accepts indeter-

---

[34] R. Haag, "Fundamental Irreversibility and the Concept of Events," *Comm. Math. Phys.* **132**, 1990: 245-251.

[35] Numerous ideas circulate with proposed resolution of irreversibity. I have earlier mentioned the ideas of Penrose and Prigogine. We note that Sakharov attributes irreversibility to CP non-conservation—the cause of the preponderance of matter over antimatter in the universe.

minism there is no reason of not including irreversibility as part of the fundamental laws of nature. We are, he argues, dealing with both induced and a spontaneous birth of facts. But, by becoming a fact, an event is irreversible and this irreversibility does not ask for an explanation in terms of statistical mechanics or the consciousness of an observer. Lastly let me note that there is an inconsistency between what physicists consider the comprehensive physical theory as embodied in thc Schrödinger equation and their "theory of measurement" which does not itself — in their eyes — belong to the theory, thus requiring a separate axiom, that of the "projection postulate."[36] This inconsistency disturbs the reductionist belief (heuristic) in the unity of science as Scharf (1987) notes. Many lately, have attempted to resolve this difficulty with the introduction of "decoherence"[37] — if I correctly understand it, an inescapable process for the mixing of phases at a microscopic event space as it interacts with the extant macrostructures. This approach introduces irreversibility and the arrow of time right where events occur: in transitions at the microscopic quantum level. The effort suggests that indeed there is an agonizing problem to be resolved. As I see it, "decoherence" is just

[36] A lengthier discussion of this is presented by D.C. Scharf, "Quantum Measurement and the Program for the Unity of Science," *Phil. of Sc.*, **56**, 1989: 601- 623, where he refers to the reductionist program of Oppenheim P. and H. Putnam, ("The Unity of Science as a Working Hypothesis"). But in physics one is hard pressed to describe what is elementary and, hence, reductionism is, at the very least, a continuously updatable process that incorporates the new findings. What I have been describing in this essay addresses the related problem of building the experimental apparatus. However, things are not additive, as Scharf supposes. Each step in the knitting operation, from the first on to the last in the assembly of structures, necessitates a projection postulate.

[37] Wojciech H. Zurek , "Decoherence and the transition from quantum to classical," *Physics Today*, October 1991: 36-44. From this article one can be led to the more technical literature mostly by the same author. See also Roland Omnès, *The Interpretation of Quantum Mechanics*, Princeton NJ, Princeton University Press, 1994.

another, subtler to be sure, sleight of hand that may yet lead to a better understanding of the arrow of time. But for the moment it is an *ad hoc* introduction of an experience not yet articulated in the fashion mathematical physicists expect to have their theories formally expressed. It does not add an operational machinery to substitute for and justify the axiomatics of the projection operators of von Neumann and it does not couple with the statistical mechanical procedures that lead to the macroscopic measurements. Experiments to explore the mechanics of the collapse of the wavefunction and the cleavage between the quantum and the cognitive world are under way and one expects that clarifications may follow — perhaps even new insights are in the works. When one reads that "the decoherence phenomenon transforming this superposition into a statistical mixture was observed while it unfolded, providing a direct insight into the process at the heart of quantum measurement,"[38] one may conclude that an unabiguous support of the von Neumann conjecture is justified.

The quantum mechanical discourse is haunted by the "observer" which is anything but a well defined construct. The physicist, in the tradition of Plato, later of Kant and Descartes, unobtrusively hovers over scientific objects. He is the ubiquitous *narrator* nowhere to be seen or physically touched — a metacharacter in the terminology of structuralism.[39] According to this theory "each narrative has two

[38] Michel Brune *et al.*, "Observing the Progressive Decoherence of the 'Meter' in a Quantum Measurement," *Phys. Rev. Lett.*, 77, 24, 9 December 1996: 4887-4890. This reference along with the comment was added as I was editing the final version of this manuscript.
[39] Gérard Genette, *Narrative Discourse: An Essay in Method*, translated by J. E. Lewin, Cornell University Press, 1980, from parts of *Figures III*, Paris, Editions du Seuil, 1972: 19.

parts: a story (*histoire*), the content or chain of events (actions, happenings), plus what may be called the *existents* (characters, items of setting); and a discourse (*récit*, *discours*), that is, the expression, the means by which the content is communicated. In simple terms, the story is the *what* in a narrative that is depicted, discourse the *how*."[40] The narrator connects the story with the discourse but we see the story through his *narration* and meditations. There is also a distinction between the real author, the implied author, and the narrator with counterparts, the real reader, the implied reader, and the narratee. Surely this methodology could apply to the scientific research paper or textbook and could help to clarify the process of exposition without undermining the established order. One can imagine a fruitful and instructive analysis of the scientific narrators (say, the austere Dirac or the more entertaining Feynman) in terms of structuralist narrative theory: an excellent idea for a term paper that may parallel the processes of scientific narration with some aspects of fiction writing; after all, many recent papers are speculative enough to justify this categorization. Clearly the story and the chain of events are the rate equations and their recursive solutions, while the existents consist of the environment and the Hamiltonians; the discourse is the explication of the technique of working the mathematics of the story and the interpretation.

So much for the worries and the caveats; we now return to the more practical approaches to experiment. The issue here is not to describe their quantum mechanical subtleties but rather the construction aspects of experiments. For the dis-

[40] Seymour Chatman, *Story and Discourse: Narrative Structure in Fiction and Film,* Cornell University Press 1978: 19.

cussion that follows it is the interaction space and its realization which is of concern. This is entailed in the concept of physical capital although the latter introduces the idea of value in a world where none is presumably justified except *ex post*. In this chapter, "capital" is taken in its physical connotation and with only intertemporal implications since in order to assemble an instrumentation box for measurement, a process of an atom-by-atom construction in time can lead to the macrostructure in space. Indeed, the process of measurement is a virtuality come true during the interaction of energy and matter with capital-like structures. A measurement consists of the assembly of an instrument around an interaction space where virtualities become events with signals of various sorts consistently leading to indications of happenings. But what is an instrument? An instrument can take on a number of functions with or without utilitarian outcomes. The Carnot Engine, referred to in earlier sections, is one of the former type; it is also the prototype of capital. The Stern-Gerlach (S-D) and the Low Energy Electron Diffraction (LEED) set ups, are examples of the latter. All three "experiments" have been the pillars and exemplars of modern physics; all require a "preparation" — a notoriously ambiguous idea or concept encountered in all theoretical texts. Here I describe these two last experiments from the perspective of an experimentalist.

a. The LEED Experiment

Take the example of the LEED (Low Energy Electron Diffraction) experiment which was used first to demonstrate the wave nature of the electron[41] as conjectured by de Broglie. "Preparation" in this instance pragmatically means to follow certain procedures, namely: an X-ray measurement is taken on an extant solid to determine periodicity, crystal structure, orientation, purity, etc. With these preliminaries completed, sections are cleaved or sliced in wafers, surfaces are ground, chemical cleaning and degassing follows and then the surface is positioned in a vacuum chamber facing a filament hot enough to have its electrons "boiling" off its surface. These electrons are accelerated, velocity selected and directed at the surface just positioned in the vacuum. The surface backscatters the electrons and these, in turn, impinge on a fluorescent screen which is thus illuminated with spots whose geometrical positions are indicative of the reciprocal lattice structure of the scattering surface with its two-dimensional periodicity; spots result from the electron wave interferences. What we have "prepared" is a massive packing of a vast number of atoms and molecules coalescing together in the formation of the crystal. The quasiperiodic electron density (of bound and free electrons in the solid), neutralized by the positive core charges of the atoms, is extremely high and asymmetrical at the surface. It exerts a large totality (sum) of forces on the impinging electrons, impressing on them spatial coherence and backscattering them. The scattered electron waves in-

[41] C. J. Davisson and L. H. Germer, "Scattering of Electrons by a Single Crystal of Nickel," *Nature*, **119**, 1927: 558-560.

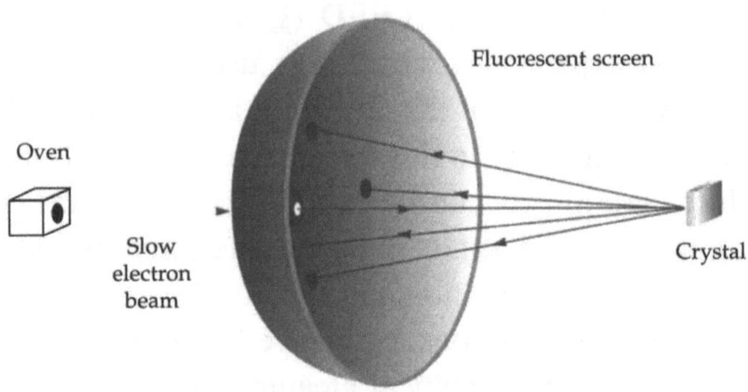

Figure V-5. The LEED setup (schematic)

terfere among themselves and the critical event appears to happen just as soon as the fluorescent screen is hit when the electron wavefunctions finally "collapse" and produce light visible to the senses. There is then a measurement sought and, subsequently, observed by that universal instrument, the retina and its extension, the brain. In this instance, the brain is assumed to be decoupled from the rest of the universe. However, once visual observation is achieved, a photograph of the event will register the information for later study.

b. The Stern-Gerlach Experiment

The other paradigmical experiment was devised by Stern and Gerlach.[42] It was conducted in 1922 as a result of curio-

[42] W. Gerlach and O. Stern, *Zeit. Phys.* **9**, 1922: 349. See also Pais (1991) for a historical perspective and D.E. Platt, "A modern analysis of the Stern-Gelach

sity, but its unexpected results provided the clue on the intrinsic moment of the electron, the spin, and opened up the path which, in time, resolved numerous fundamental quesions of the behaviour of matter and provided a simple, yet unintuitive, theoretical structure. Only later was it shown that spin is an intrinsically relativistic effect. The experiment touches on the foundations of quantum mechanics and as such deserves a discussion which goes beyond the historical account.

The experimental configuration is as follows: a narrow beam of neutral (silver) atoms traverses a small region of space which is fluxed by a *nonuniform* magnetic field (what is a magnetic field?). On exit from this region the beam is intercepted by a photographic plate. When the emulsion is fixed, one observes that the beam has split in two components whose distance is proportional to the gradient of the magnetic field.[43] Hence the mystery. What was immediately advanced was that the electrons in the atom have an up and a down magnetic moment and are, accordingly, deflected by the passage in the inhomogeneous magnetic field. What confuses the issue is that the experiment refers to atoms of silver — a red herring of sorts — while the theory refers to electrons. Do free electrons behave similarly? The answer is that the silver atoms are not essential;[44] being

---

experiment," *Am. J. Phys.* **60** (4), 1922: 306-308. This analysis, however, indicates that we may derive explicitely from the dynamics the outcome of an experiment, without recourse to any other consideration (the postulate on measurement) See reductionism etc.

[43] The force being $\mathbf{F} = -\nabla E$ where E is energy.

[44] H. Dehmelt, "Experiments on the Structure of an Individual Elementary Particle," *Science*, **247**, 1990: 539-545, and earlier work where it is shown that the spin magnetic moment of a free electron can be measured (an idea earlier disputed by Pauli and Bohr). However the techniques required are much more

massive by comparison to the electrons, however, they move slowly and are easily handled experimentally. They act as slowly moving carriers of the electrons which trace their trajectory leaving a macroscopic observable; it is through the nickel atoms that the target emulsion is chemically activated and the effect observed. On development of the plate, the silver atoms leave the traces of the atom trajectories. In a very roundabout fashion we thus infer the existence of a "magnet-like" electron. Once again we encounter the idea of observation on a photographic plate. The microscopic events, unlike their traces, are not observable.

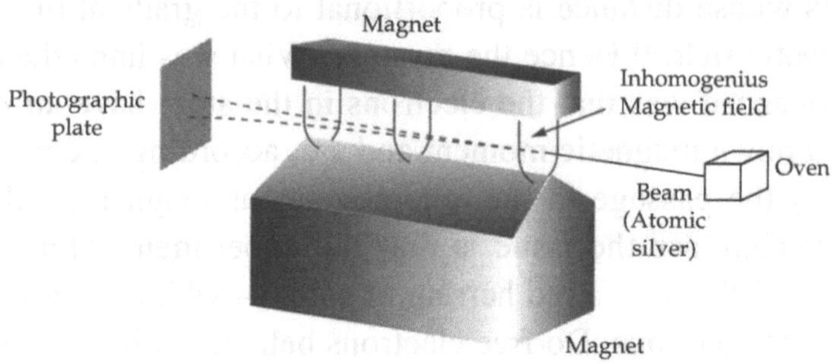

Figure V-6. The Stern-Gerlach set up

"Preparation" in this instance consists in collimating a beam of silver atoms traversing an "external magnetic field." In the theoretical texts this latter is included in the

---

exact and dependent on later electornic and microwave technology. Polarized electrons have been seen emitted by Gallium Arsenide (GaAs) chips .

"external" Hamiltonian H_{ext}. It is taken to be both macroscopic and given; it can be the result of spin-originating magnetism, of normal or supeconducting Amperian currents. The totality of the magnetic photons which are virtually emitted from the massive collection of spins or currents (the "external" magnetic field) exert a "force" that is just a whisper on atoms in the interaction space. Intermediated by the "orbital" so-called electrons, with which the magnetic photons interact, the atoms are dragged away from their linear path. "Preparation" appears to imply not only to prepare the crystal or the particle but, also, to condition the interaction space for it is there that, at interaction time, the massive exchange of photons takes place.

What then is an experiment? Once again the question remains as to what exactly is meant by experiment. And this is as good an opportunity to dwell on the subject. Whatever the mechanism, measurement is itself the result of a "brick-by-brick" assembly of matter. In this manner I want to impress on the reader the importance of structures and their formation, an engineering process which in the world of theory and of experimentation has lost its inherently economic value. "Experiments" apparently consist of interactions of single or few particles with a *massive* assembly of coherently acting "particles." Each of these structural elements of a measurement is itself the product of earlier formations. For a more theoretical approach we may remind ourselves of the introduction by Schwinger of nonoperator (i.e. classical) sources and fields to replace the local field operators. These sources symbolize the interventions that constitute measurements of the physical system[45] and of the origins of the concept of response functions for macroscopic systems. In our approach here they replace the summation of all past history in the structuring of these sources and fields.

In addition to the above "deliberate" experiments, there are also "stochastic" and serendipitous experiments which happen independently of an observer-experimenter. The construction of a critical assembly for nuclear reactions is a deliberate experiment; yet the Oklo reactor mentioned in ftn. 33 indicates that, in nature, a stochastic conjunction of events has led to a chain-reacting fission structure. Nature has provided natural magnets, electrostatic devices (amber) and effects, plain or birefringent crystals, not to mention the thermal cycling engine, that is the atmosphere of the earth which is driven by the sun's energy. Molecular replication belongs to the same category and, surely, in its evolutionary development, provides the basic interactive instrument for the study of nature itself through the deliberate design of experiments. In the chapter on biology the objective was to look into the formation of reproducing structures. Clearly, in terms of the theory of measurement, the formation of a biological cell is a "measurement" as the completed structure interacts with particles and energy inside and outside its volume to maintain and reproduce itself. The question then is not "whether", but rather "how", to connect what is the measurement apparatus in the language of the theory of measurement with the object to be measured i.e. the new particles that will contribute to reproduction of the measur-

[45] For comments on this important point see Cao and Schweber (1993: 50, ftn. 29).

ing equipment. In biology we are then faced with a problem physicists have avoided for too long.

One could go on with the recitation of many more "experimental preparations," all methodologically variants to the ones just described. But it is, perhaps, more practical at this moment to propose tentatively some principles of measurement:

First, "measurement" is a continuously developing and improving process which depends on the techniques of assembly of capital-like structures that define the interaction spaces of the experiments and materialize virtualities.

Second, for a measurement to proceed, there must be a non-equilibrium to allow for the coherent (net) flows of heat, electrons or other indicators of events.

Third, "every measuring device is an aggregate of *quantons* with macroproperties." This includes the biological structures and the brain.

The first two principles have been elaborated throughout this essay. The third, however, needs further commentary. It is the content of Bunge's 10-th axiom of Quantum Mechanics and has, I think, important conceptual content. It implies that every system consisting of a quanton and a measuring device is a physical sum $\sigma + \sigma'$; and this, contrary to older theories which would have us believe that σ' only represents the environment of its partner σ.[46] The interesting

[46] See p. 277 and 278 of Bunge (1967). *Quanton* is a word used to represent the types of 'particles' belonging to the realm of (i.e. they are the references of) the quantum mechanical world.

processes in the physical (and the biological) world do happen at interfaces. The "cleavage" between the homogeneity of infinity and the finite slices of the world is the beginning of physics and of its comprehension; the essence of measurement,[47] of capital formation,[48] and of dialectics. As to thermodynamics and statistical mechanics it appears that they have no seeming usefulness in a homogeneous world or even a homogeneously fluctuating infinite world. The cleavage is a necessary division and it is the precursor of the invention of capital. The basic questions are then transferred to the issue of why did the cleavage ever appear. Is it the result of some fundamental original imbalance? An

---

[47] We note that initiation of an experiment itself is a difficult theoretical problem. For instance, how does one describe the triggering of a laser or the initiation of criticality in a nuclear reactor? See the comments in chapter III, ftn 33, which are quite relevant to the issue. The literature on quantum mechanical measurements is so speculative that it is difficult to even suggest a view which could guide the reader. So we mention here an article by H. Primas and U. Müller-Herold, (in Rice 1978), which can provide some illumination, references, and a critique. The point is that in classical mechanics one assumes a sharply defined particle and trajectory and in the multitude the details are lost so that ensemble averages have to be dealt with. In the quantum case we begin with a built-in probabilistic behaviour of each 'particle' and then once more have to deal with the statistics as in the classical case. A useful discussion is presented by D. Deutsch, "Quantum Theory as a Universal Physical Theory," *Intern. J. of Theoretical Phys.*, **24**, 1, 1985: 1-41. There, he questions the notion of the ensemble, and disputes the accepted interpretations of quantum mechanical measurements. He suggests that entropy is not "necessarily an increasing function of time: there can be no such function in a time reversal invariant theory such as quantum theory." Entropy "tends to increase whenever hitherto kinematically independent subsystems begin to interact." The idea is that the world has evolved various parts of it differently. The interaction among these parts is the mechanism that makes for the observed entropy increases.

[48] To my knowledge there has been another association of 'capital' with a physical problem. See F.J. Tipler, "Some thoughts on the analogy between quantum mechanics and the Austrian theory of capital," Draft Preprint, Center for Theoretical Physics, University of Texas, Austin, and Depatment of Physics and Mathematics, Tulane University, New Orleans, 1981. I would like to thank professor Tipler for sending me the reprint. Tipler considers F. Hayek's definition of capital (an extension of that originally given by Böhm-Bawerk) as analogous to the "Feynman sum over histories," but more general because of its dependence on all other histories.

ever so small imbalance of matter over antimatter?[49] If so, then, in the end, we might still state that the "2nd law of thermodynamics is a fundamental law of nature, deriving from an original cosmological imbalance." We can thus be accused of summoning the Maxwellian Demons to bring about the $\sigma + \sigma'$ cleavage which resulted (by necessity but not sufficiency) into the formation of the first capital, and thus, provided the germs for its reproduction.

Μέλη πεποίηκ' εἰς τάς Νεφελοκοκκυγίας (917)[50]

---

[49] In "Singularities and time-asymmetry," p. 787 of *General relativity*, S.W. Hawking and W. Israel, eds. Cambridge University Press, 1979. Penrose makes the intriguing proposal that one should impose on the initial singularity a time-asymmetrical initial condition - namely, the vanishing of the Weyl tensor - and that this is related to the second law of thermodynamics via the field's contribution to the total entropy density of the universe.

[50] I have composed songs to honor Nephelokokkygia.

As an introduction to the next chapter:

$$\frac{h}{\mathrm{i}}\frac{\partial \psi}{\partial t} + \mathrm{H}\psi = 0$$

The Beginning (Erwin Schrödinger)

Il n'y a dans le visible que des ruines de l'esprit

Merleau-Ponty*

**Le Monde* 02.05.1981. The juxtaposition of Schrödinger's equation, an equation which in its more general form describes all happenings in this world, is quite fitting. Indeed, Schrödinger's equation describes transitions from state to state; yet rarely in physics one is interested in the "ruines," as it were, which are left behind.

CHAPTER VI

SUMMARY AND CONCLUSIONS

THE QUESTION THAT MOTIVATED THIS RESEARCH was the study of the origin of capital. This naturally led to biological and thermodynamic structures and, in turn, to the substancce of thermodynamics itself. Is thermodynamics a field on a par with, say, electrodynamics? With quantum mechanics? Apparently not, since thermodynamics is the result of utilitarian concerns; to be understood it needs the concept of value. Value derives because a biological entity whether by design or evolution must shed information (data) if it is to survive. Thermodynamics is *par excellence* the field that does this successfully. It is the epitome of the concept of scarcity and with its engines and processors it provides the more lucid examples of capital structures. This justified the lengthy critique of thermodynamics. At the same time these structures have to be assembled. This gave me the opportunity to explore the physical processes of coalescence of matter and of the formation of structures in order to acquire a sense of what is known in order to discuss experiment.

To be familiar with the state of current science in its totality appears to be an impossible task. Somehow, the mind manages to isolate features of this totality and concentrate on them, decoupling, as it were, the peripheral vision which distracts, and isolating systems and disciplines weakly interacting with their neighbours. Thus, in setting a problem in one's thoughts it becomes necessary to identify, if at all possible, the existence of the peripheral issues and, yet, to neglect them. This has been the approach followed by professionals in the fields touched in this essay. Here I have attempted to identify and then follow a common thread among disciplines. To remain current even in the frontiers of one's own field is arduous; very few manage. Indeed, any "text is a sluggish machine requiring of the reader strenuous effort."[1] One then depends on the judgement of able and imaginative concentrators: appreciation of their limitations and their reasons and biases is essential while misinterpretations remain one's own.

In the world of economics, *capital* triggers very specific responses although it is a concept as elusive to catch as mercury and as dangerous to handle as a venomous snake, the venom coming mostly from its theoreticians. Capital touches on deep emotional problems of distribution and control of decisions and as such it has preoccupied Marx, the neomarxists as well as the neoclassical economists. It is, after all, *the* focal point of economics. With this prevailing outlook and professional emphasis it is difficult to introduce yet another concern: the origins of capital. The divide is between the conscious effort for planning investment and the aleatory evolution of capital which is observed in the biological world. The latter aspect of the problem is my concern. In reviewing some fundamental problems in the economics of capital for-

---

[1] U. Eco, *Lector in Fabula,* Bompiani, Milan 1979.

mation and in statistical mechanics I have come to the conclusion that capital and its historicity is a universal construct, basic in the search for the origin of life, in the reconciliation of thermodynamics with statistical mechanics, and the economics of value. As noted, however, this construct has acquired the status of a concept — an undefinable idea — because in economics the physical problems are lost in the process of allocation of resources which are valuated on the merits of their scarcity rather than on their physical performance characteristics.

To begin with, I have adopted in Chapter II Keynes's basic formulation to the effect that "it is by reason of the existence of *durable equipment* [capital] that the economic future is linked to the present."[2] The stress on the physicalization of the intertemporal virtual process suggested that we look deeper into the origins of capital. However, the search into the economic literature did not provide any clues except a few stylized biological examples of accumulation and reproduction (seeds) that illustrate the basics of the economic "equation of motion." This led to the study of biological structures which have the characteristics of stochastic evolutionary processes only to conclude that in whatever level, capital (the structure) and energy are the important variables in F(**K**, **E**, t), what we call the Universal Production Function. Yet "nothing happens in the world, not a flutter of an eyelid, not a flicker of a thought, without some redistribution of microphysical states."[3] This redistribution cannot be a random event. Rather it must occur in a structure, itself the outcome of earlier redistributions. Physical processes and

---

[2] *General Theory*, p. 146.
[3] W.V. Quine, "Otherwordly," a book review in *The New York Review of Books*, 23 Nov. 1978.

observations cannot obtain unless we have these structures in place. Following the prevailing dogma, structures develop out of a concatenation of wavefunction collapses. By looking into the biological realm one probes into the foundations of biology and physics and there recognizes that the interplay of capital with the redistribution of microstates is the key to *happenings*, to observation and to replication. The origin of capital and life are interconnected although the question in biology which relates to the identification of the mechanism that produced the first replication is still open. How a cell replicates is by now almost comprehensible; so that the real question is how the cell "invented" itself. While the sufficient conditions are still unknown, the necessary ones, I suggest, may be found in the interaction of capital structures with energy.

Recognition of these mechanisms does not in itself lead to revolutionary change of view in physics or biology and it does not provide any new algorithms for calculation or a contribution to scientific reductionism — a valued objective and procedure in the making of good science. But, as I noted in the Introduction (p. 8), not everything worth establishing can be translated into the algorithmic vocabulary of physics.[4] Often, an incisive outlook or perception will open important research paths. Such has been the case of Darwin's theory of evolution and most probably will be the case of the theory of consciousness. Problem setting and solutions can take various forms. Some are based on superb mathematical architecture (Dirac's work is an exemplar here) but some are the result of deep understanding and intuition. For instance, Lan-

---

[4] *Ibid.* form which I paraphrase and use "algorithmical" for Quine's "technical."

dau's celebrated work[5] has much of this characteristic. His were solutions without derivations that took years to prove them valid. Their inspiration often was based on catching a key characteristic of a phenomenon and building a heuristic mathematical description. When it comes to consciousness one may expect this type of inspired early approaches. For to understand the electronics of neurons may not add up to consciousness. Similarly, the formation and accumulation of capital, in all its different ramifications, is not a process with any fundamental physical content, although structurally, its assembly and use can be described solely in physical terms. On the other hand, as I discuss in chapter V, without capital-like constructs there could be no observation. And the enigma remains as to the formation of the first (coherent) capital out of the *néant* whatever that may mean. Maxwellian demons may still be invoked in the biological world which, without intelligent guidance or clever manipulation, are responsible for the construction of assemblies that mediate energy transactions.

Assuming then that the origin of this mechanism can be established, and in view of the fact that the possibility of the coalescence and stability of matter (see p. 119) is inherent in the quantum mechanical rules, one conclusion can be reached: the legitimation of thermodynamics can result from the study of the interactions of capital and energy; the capital structures are assembled *brick-by-brick*, develop into macroscopic bodies massively interacting, in turn, with the elementary units of matter (the "quantons"). It is the contributions of work, i.e. the energy required to add or subtract a

---

[5] For a taste of Landau's style refer to D. ter Haar, *Men of Physics: L.D. Landau*, Volumes 1 and 2, Pergamon Press, 1965.

particle from a body, that is responsible for the assembly of capital-like structures, while the massive boundaries of the capital determine the summation of the interactions which take place within. All this suggests the following program for the cleansing of thermodynamics and of statistical mechanics:

1. Accept the necessity of *capital* (physical, finite structures) for the mediation of energy transactions.

2. Introduce the intertemporal concepts of capital formation and the ensuing constraints in thermodynamic applications thus intimately connecting physics with thermoeconomics (see Appendix 3).

Following this methodology, the 2nd law will emerge in the context of energy transactions mediated by capital structures which are needed to negotiate with the future in order to obtain *work* now (discounting!) In this manner, an economic foundation can be given to the 2nd law of thermodynamics, the law often and succintly referred to as the most economic of all laws. Note, however, that the "law" in our approach is only a *constraining* rule on the first law of energy conservation; it results from the utility to assign a premium to *work*. This rule comes into play when energy-capital interactions occur to give a measure of work forgone. Recall that Planck considers the 2nd law as a "fundamental law of nature" independent of the observer. But in the present approach it is the observer and the necessity — utility — for observation that make the 2nd law a useful statement on constraints. These considerations are of no interest to a non-dialectical system yet they are the principal determinants of scarcity.

Scarcity as such is not a physical concept; it only plays on physical rules. In fact scarcity is the essential ingredient in the making of history. Or as Sartre wrote succintly:

> Mais dire que notre Histoire est Histoire des hommes ou dire qu'elle née et qu'elle se developpe dans un cadre permanent d'un champs de tension engendré par la rareté, c'est tout un...
> La rareté fonde la possibilité de l'histoire humaine et non sa réalité,[6]

a seminal idea which may become a guideline for various disciplines. Economics accepts the reality of scarcity as a given and prescribes substitution and innovation as techniques for obviating it. That *scarcity* is the fountain of history, and here that includes the origins of biological existence, is a striking statement rich in philosophical and practical implications.

To end this essay I juxtapose a philosopher's view of the physical world with that of a scientist. Sartre (probably inspired by the "Copenhagen School" of Bohr) writes that "the only theory of knowledge which can be valid today is the one which is founded on that truth of microphysics: the experimenter is a part of the experimental system. This is the only position which allows us to get rid of all idealist illusion, the only one which shows the real man in the midst of the real world ... the revelation of a situation is effected in and through the *praxis* which changes it."[7] This is perfectly

---

[6] "La matière travaillée" *CDR*, Sartre p. 200.
[7] From Sartre's *Méthode*; Barnes' translation, p. 25. An interesting passage which merges Bohr's theory of measurement (plagiarized? without giving credit) with Marx's dialectic. He accepts the "truth of microphysics," a truth

compatible with Dyson's (1979) views on the remote future; and it is no different from the views of Wheeler (1979) who depicts "the Universe (with a big U])... as a self-excited circuit growing to give rise to observer-participancy which in turn gives tangible reality to the Universe"[8] and who sketches the U with an eye for a serif! — a noteworthy development away from Planck's "idealist illusions" discussed earlier. It is a model related to S. Beckett's[9] play *Not I*; a mouth can be perceived on stage, a logodiarrhic box, a structuralist symbol existentially searching for an "I". Our function in this world might as well be summed up thus: *Gossipo, Ergo Sum.*

But what is the mechanism which allows observer-participancy and gives tangible reality to the Universe and how does the virtual Cartesian observer operate? Scientists feel

---

one supposes, that can be reached asymptotically with research. This truth, however, being non-dialectical is unalterable with participation. The dialectical derives from the variety of possible outcomes which the truth of microphysics permit. Sartre's Marxism may not be all that pure as one may argue on reading (*Méthode* p. 25) that "the physicist's hypothesis, before it has been confirmed by experimentation, is also an interpretation of experience; it rejects empiricism simply because it is mute. But the constitutive schema of his hypothesis is universalizing, not totalizing. The Marxist approaches the historical process with universalizing and totalizing schemata..." One also suspects that Sartre may have been influenced by Schumpeter's early work on the subject of equilibrium.

8 J.A. Wheeler, *Frontiers in Time*, in Corso 72, School of Physics *Enrico Fermi*, G. Toraldo di Francia, ed. Amsterdam 1979. This places Wheeler in the category of Dialectics in the philosophical table of Chapter. II, 3. Wheeler recognizes the self-observation of the Universe through the biological instrumentation. For more see *Int. J. Theor. Phys.* **21**, 1982, Papers from Conference on Computation. Also Wheeler, J.A., "The Computer and the Universe," pp 557-600 which relates to the theory of measurement discussed in Chapter IV. 6, and J.A. Wheeler, "On recognizing 'law without law'" *Am. J. Phys.* **51** (5), 1983: 398-404, which has a bearing on our views for the assembly of structures.

9 S. Beckett, *Pas Moi*, Les Éditions de Minuit, 1975; this is Beckett's deeper play.

that the time is now ripe to tackle these problem feeling that the understanding of the physics of the brain will permit the resolution of the "mystery" of thought and of consciousness in particular. Clerk Maxwell's "materialist" view that "... the only laws of matter are those which our minds must fabricate, and" that "the only laws of mind are fabricated by matter,"[10] is the one adopted here expecting that, along with evolutionary considerations, it will lead to progress in this search.

---

[10] Quoted in Tolstoy (1982: 77). Quite a different view from that adopted by Planck.

Appendices

1. On Thermodynamics and Statistical Mechanics
2. Availability
3. Thermoeconomics
4. Tractatus de Signis. Utilitarian Concerns and Scarcity
5. Prolegomena on Measurement

Why the Appendices? They are not meant to be completions of arguments begun in the main text. Rather they are "leafings" through an extensive literature. A researcher scans enormous amounts of information and decides where to focus his attention; and even then he is diffident about his sources. To replicate the sequence of proofs in any article is an extremely arduous enterprise. He must decide on what is essential and then reject, reject, reject or as Henry James put it, "intensely to consult and intesely to ignore." What I attempt to do here is to make public (not to *publish*) a small — and a very small at that — set of notes from my perigrinations. Items, perhaps, that could lead to a research paper or the exploration of an idea.

Most research (as with most theses) is safe in that it is done on established, working paradigms and techniques; nevertheless Appendices four and five indicate scientific areas that are open to further research and argument. But as noted repeatedly science has progressed without the benefit of perfections in the foundations.

1. On Thermodynamics and Statistical Mechanics

In this appendix I go through some sketches that may support the heuristics in the text. The objective is to show the flow of procedure rather than the detailed proofs. The legacy is that of Gibbs rather than that of classical phenomenological thermodynamics.

The Gibbs ansatz

The *Gibbs ansatz* refers to an equilibrium canonical distribution function $f_N(p,q)$ given by

$$f_N(p,q) = Z(\beta)^{-1} e^{-\beta H(p,q)}; \quad Z(\beta) = \int d\Gamma e^{-\beta H(p,q)}$$

where $\beta = 1/kT$, T being the (absolute) temperature. However our interest is moving away from equilibrium by applying an external force characterized by the [parameters a_s and its variations δa_s. The time-dependent distribution function $f_N^0(p,q,t) = f_N^0(p,q,t_0) + \delta f_N^0$ is now written in the form

$$f_N^0(p,q,t) = \exp\left\{\frac{F(t) - H(p,q,a_s(t))}{kT}\right\}$$

$$\text{with } e^{-\beta F} = Z^0 = \int d\Gamma e^{-\beta H(p,q\ldots)}$$

since $f^0_N(p,q,t) = f^0_N(p,q,t_0) + \delta f^0_N$ integrates to 1 its variation is zero and therefore with a bar over a quantity signifying an average over a whole system the number of particles remaining fixed.

$$\delta \int d\Gamma e^{\frac{F-H}{kT}} = \delta\left(\overline{\frac{F-H}{kT}}\right)$$

hence

$$\delta F = \left(F - \overline{H}\right)\frac{\delta T}{T} + \overline{(\delta H)} = \left(F - \overline{H}\right)\frac{\delta T}{T} + \sum \overline{\frac{\partial H}{\partial a_i}}\delta a_i$$

The expression following the summation Σ is what in engineering is called "shaft work" or "external" work i.e. $\delta \mathcal{W}$.

The reason for going through this exercise was to indicate the distance between the Gibbs ansatz and the formulae needed for phenomenological thermodynamics in which the inability to give physical content to heat results into some peculiar mathematics (incomplete differentials which inordinately confuse the student of the field). Here, however, we follow Gibbs (1902) in his statistical definition of work. heat and entropy. If the system is driven by some external ("generalized") forces that each leads to a displacement δa_i, the mechanical work *received* by the system is

$$\delta H = \sum_i (\partial H/\partial a_i)\delta a_i.$$

By definition the *internal energy* E and the thermodynamic *work* $\delta \mathcal{W}$ received during a quasistatic change are given respectively by $\langle H \rangle$ and $\langle \delta H \rangle$ where H is the equilibrium

Hamiltonian. The change of the internal energy δE is in general different from average received work $\delta\langle H\rangle$. In fact the difference $\delta\langle H\rangle - \langle\delta H\rangle = \delta Q$, the heat received from the thermostat. Using 4x we find that

$$\delta\left(\frac{F-\langle H\rangle}{kT}\right) = \frac{\langle\delta H\rangle - \delta\langle H\rangle}{kT} = \frac{1}{T}\delta Q = \delta S$$

where S is the entropy. Since

$$\log f_N^0 = \frac{F-H}{kT}, \quad \text{its average is}$$

$$\langle \log f_N^0\rangle = \frac{F-\langle H\rangle}{kT} = -S = k\int_{\Gamma^0} d\Gamma f_N^0 \log f_N^0.$$

From these follow the expressions

$$\delta F = S\delta T + \langle\delta H\rangle \text{ and } \delta\mathcal{W} = \langle\delta H\rangle + T\delta S = \langle\delta H\rangle + \delta Q.$$

These are the basic equations as derived using the Gibbs ansatz. However, to go beyond these equations that describe a system permeated by the external force and into the so called *grand canonical ensemble* a few modifications must be introduced. One must specify a closed volume in interaction with its surroundings giving it a shape resembling the structure of a biological cell. But then a subtler and more general derivation of the various quantities is essential. For the purpose we must allow for the exchange and the production/annihilation of particles. The more general method, the *grand canonical ensemble* (GCE), has been devised by Gibbs (1902) and independently introduced into engineering in the nineteen thirties. The GCE consists of a partial volume V_I within a much larger one V_{II} ($V = V_I + V_{II}$) described by some arbitrary statistical state $f_N(\boldsymbol{p},\boldsymbol{q},t)$ and circumscribed by a real or fictitious surface; in the former case, V_I refers to a local measurement domain (in the hydrodynamical state); in the latter it may refer to a diathermal or isolating membrane (for energy transfers), to a permeable, semi-permeable or, impermeable membrane through which an exchange of particles may be effected. The Hamiltonian of the total system can be split into two separate ones H_I and H_{II} if the interactions between particles across the border can be neglected. Then one can write

$$f_I(p_1,\cdots q_{N_I};N_I) = \exp(1/kT)[F(N,V) - F(N_{II},V_I) - H_I(p_1,\cdots q_{N_I},N_I)]$$

thus extending the above relations. However, when $V_I << V$ and $N_I << N$, we can write

$$F(N,V) - F(N - N_I, V - V_I) \cong N_I\left(\frac{\partial F}{\partial N}\right)_V - V_I\left(\frac{\partial F}{\partial V}\right)_N = N_I\mu + \Omega_I$$

where $\mu = (\partial F/\partial N)_V$ is the definition of the *chemical potential* of the system and $\Omega_I = - PV_I$, where P is the pressure of the total system — a constant when there are no external forces. Thus finally,

$$f_I(p_1,\cdots q_{N_I};N_I) = \exp\{\exp(1/kT)[N_I\mu + \Omega_I - H_I]\}$$

The entropy then takes the form

$$S_I = -\frac{\overline{\overline{N}}_I\mu + \Omega_I - \overline{\overline{H}}_I}{T}$$

$$= -k\sum_{n=0}^{N}\int_{\Gamma^0(n,V_I)} d\Gamma^0 \; f_I(\mathbf{p},\mathbf{q},n)\log f_I(\mathbf{p},\mathbf{q},t)$$

where the double bar signifies an average taken over a local statistical state while a single bar or the brackets <> refer to the whole system with a fixed number of particles. Once these expressions are secured then a generalization should be introduced by changing F to G with G = F + pV +μN

$$\delta\Omega = \delta\mathcal{W} + \delta Q + \mu dN$$

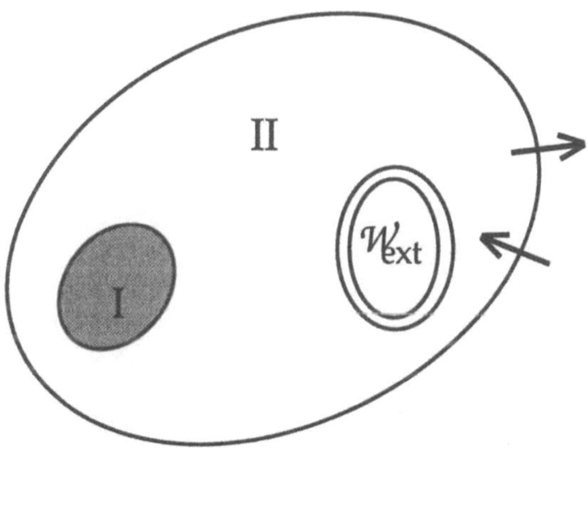

Figure A-1. The Grand Canonical Ensemble. Region II

The legitimation of phenomenological thermodynamics can be accomplished within the Gibbs ansatz itself delphic on the origins of irreversibility.

$$\delta H = \sum_s \frac{\partial H}{\partial a_s}\delta a_s = \sum \mathbf{F}_s \cdot \delta \mathbf{a}_s \quad ;$$

this is an expression for the interplay of the external parameter $\mathbf{a}_s$ of direct macroscopic import — presumably connected with the design of the equipment — with the internal properties of microscopic, mechanical origin.

As I argue in a subsequent section, the key interest (if not justification or even obsession) of thermodynamics is the production of work; as a consequence various measures have been developed to quantify it. This brings us to the next section on Availability.

2. AVAILABILITY

As we have seen, the state functions include the Helmoltz free energy F and the Gibbs free energy G; to these now we must add the engineer's availability. The appropriate state function depends upon whether the transformation is performed at constant temperature, F, and parametrized by b and n; or at constant pressure and temperature, G and parametrized by β and μ; or in an environment at a fixed temperature and pressure that differs from that of the system performing work, A. It is not a bona fide state function but rather a quantity that permits the reference of energy transactions to a base state (the environment). The analysis of A is important in engineering applications but its derivation is based on heuristics rather than an analysis along the lines this far described. One possible way is to follow the grand canonical route by writing

$$S_I T = -(\overline{\overline{N}}_I \mu + \Omega_I - \overline{\overline{H}}_I) \text{ or,}$$

$$\overline{\overline{H}}_I = S_I T + \overline{\overline{N}}_I \mu + \Omega_I = S_I T + \overline{\overline{N}}_I \mu - PV_I$$

On keeping temperature and pressure constant we can write

$$\delta \overline{\overline{H}}_I = T\delta S_I + \mu \delta \overline{\overline{N}}_I - P\delta V_I$$

Were we now to include a base state with T_0, μ_0, P_0 and other things equal we have

$$\delta\overline{\overline{H}}_{I0} = T_0\delta S_I + \mu_0\delta\overline{\overline{N}}_I - P_0\delta V_I \text{ and}$$

$$\Delta\overline{\overline{H}}' = \delta A = (T - T_0)\delta S_I + (\mu - \mu_0)\delta\overline{\overline{N}}_I - (P - P_0)\delta\overline{\overline{V}}_I$$

the quantity A is known as *availability* .[1] Let us now follow how (theoretical) engineers analyze the problem of work. Within an extended medium (II) having the characteristics V_{II}, T_{II},P_{II}, S_{II} (for volume, temperature, pressure and entropy respectively) we embed two small systems (I & $\mathcal{W}_{ext}$); (I) has the characteristics V_I, T_I, I_I, S_I and is the system under study while $\mathcal{W}_{ext}$ is thermally isolated from the medium. If, while changing state, I does work on $\mathcal{W}_{ext}$ then on getting back to the original state $\mathcal{W}_{ext}$ does work on I. The maximum work in the first situation will require the minimum (and equal in absolute terms)[2] work in the reverse situation. Thus $|W|_{max} = W_{min}$. We now sum up the various ways work done on I by $\mathcal{W}_{ext}$ and the medium II:

PROCESS	QUANTITY
Work done by $\mathcal{W}_{ext}$ on I	W
work done on II by medium I. (This entails changing the volume of II by applying pressure)	$P_{II}\,dV_{II}$,
Heat given up by II	$-dQ = -T_{II}\,dS_{II}$

[1] See: A.B. Pippard, *The Elements of Classical Thermodynamics,* Cambridge University Press, 1957: 100. L.D. Landau, and E.M. Lifshitz, *Statistical Physics*, Pergamon 1958, and J. Kennan, *Thermodynamics,* Ch.XVII on "Availability", MIT Press 1941, paperback 1970.

[2] The convention is to count as negative the work (W) done *by* a system. W < 0; |W|, the absolute value is thus always positive.

Summing up the change in energy dE is

$$dE = W + P_{II} dV_{II} - T_{II} dS_{II}$$

Nevertheless the total volume remains the same so that $dV_I + dV_{II} = 0$. Surely entropy increased in the process so that at a minimum we have $dS_1 + dS_2 \geq 0$ so that

$$dE = W - P_{II} dV_I - T_{II} dS_{II}$$

or

$$W_{min} = dE + P_{II} dV_I + T_{II} dS_{II} \approx dE + P_{II} dV_I - T_{II} dS_I = dA$$

where

$$A = E + P_{II} V_I - T_{II} S_I$$

a function known as *availability* or *exergy*.[3]

[3] It differs in both meaning and form from the Free Energy $F = pV - ST + \varepsilon$ of Helmholtz.

3. THERMOECONOMICS

An engineering description of a real thermal process can get quite involved if it is to become a credible assessment of the costs and if it includes algorithms for optimizations. A suggestive accounting expression for such a calculation is in connection with a desalination project which is instructive because, unlike the operation of an internal combustion engine, say, it is very capital intensive.[4] It takes the accounting form:

$$\sum_{b} C_{\lambda}^{*}\dot{E}_{b}^{\lambda} + C_{r}^{\lambda} + \sum_{b} c_{M,b}\dot{M}_{b} = 0.$$

In words: (the forms of the various expressions of exergy crossing a boundary zone) + (the amortized costs of capital equipment) + (the costs of the material streams) = 0.

In the formula r denotes a zone in a process, C* the unit costs of exergy (the fuel or heat from a nuclear reactor), λ is an index covering all the processess involved in the operation: heat, work, fluid flow, heat flux, convection, fluid chemical convection, diffusion; $\dot{E}$ and $\dot{M}$ refer to exergy and material fluxes respectively. Note that the cost of any engineering process may be reckoned as a fixed plus an operating cost. To pay for the fixed cost it is imperative that the rate of operation be fast enough to produce reasonable

[4] See R. B. Evans, G. L. Crellin, and M. Tribus, "Thermoeconomic Considerations of Sea Water Demineralization," in K. S. Spiegler, ed., *Principles of Desalination*, Academic Press 1966.

amounts of valuable products. A faster rate of operation requires greater dissipation of exergy which is given by

$$T\dot{S} = \sum \text{fluxes} \times \text{potential drops}$$

To pay for the fixed costs, an engineering system must be designed to use considerable more energy per unit product than the minimum required by theory. In the case of a desalination facility, for instance, the cost is given by

$$y = \frac{C_1}{A} + c_A A + C_3$$

where $C_1 = cTQ^2 / T_H T_C U\dot{F}$, A is the area of the heat exchangers,. with $\dot{F}$ being the fresh water supply rate. From this formula various tradeoffs can be calculated. What stands out, however, is that the surface area and its cost, c_A, is the key variable in the calculations because it is through the surfaces, that are heat conducting or insulating, that the systems interact with their environment; these surfaces represent the bulk of the cost of capital. And this is the point we wished to make. Whether in a biological or an engineering process facility, a tradeoff is operating. In engineering, calculations are made based on the physics of the situation and the market value of the resources. In the evolution of the biological structures it is the adaptation of a structure to its environment and its effectiveness to draw resources from it in competition with other alternative structures that matters. A biological processor that has many characteristics of the desalination plant is the intestine with its high surface to volume ratio as noted in the main text.

4. TRACTATUS DE SIGNIS:[5] UTILITARIAN CONCERNS AND SCARCITY

The general concept of information or semeiotike has yet to be described with any degree of precision; efforts in this direction have not been conclusive. The reason is that the links between physical events and mental comprehension (assuming the two are different but, as a first dualistic approximation, weakly coupled) are complex and still not well understood. In this connection it is, perhaps, useful to bring to mind Locke's[6] three categories: *physica* or natural philosophy (physics in the modern terminology), *practica* or ethics (and this includes economics) and, *semeiotike* or the "Doctrine of Signs." According to Locke the subject of the latter is to "consider the nature of signs the mind makes use for the understanding of things, or conveying its knowledge to others." That includes the decision-making activities of the mind. Accordingly, information theory and its practice is seen here as constituting a very well defined subgroup of semeiotike. From this point of view, however, which is that of the communications engineer, the semeiotics are tied to the words themselves and not to their content. This is very well exemplified in the work of Claude C. Shannon[7] which is devoid of any explicitely stated utilitar-

---

[5] The title has been borrowed from John N. Deeley,with R. A. Powell, *John of Saint Thomas: 1589-1644. Tractatus de Signis: the Semeiotic of John Poinsot*, Berkeley, 1985.

[6] See his Essay *Concerning Human Understanding,* Book IV, Cchapter XXI, p. 460.

[7] Claude Shannon, "The Mathematical Theory of Communication," *Bell System Techical Journal,* 27, 1948: 379-423.

ian concerns. As Koopmans,[8] a physicist turned economist, has remarked, "if the cost of information is to be balanced against its contribution to decisions, a more general method for quantifying the several relevant aspects of information is needed." This program has yet to be worked out. Other economists[9] have attempted to identify an economic content in information. Arrow concluded that a utility function in the form of a logarithm implies that information has no rewards and vice versa. They compare values of a world with, and one without, communication channels and they thus rederive Shannon's logarithmic expression for "entropy." The formal similarities with the looks of the statistical mechanical entropy has induced considerable speculation as to the connections between information theory and statistical mechanics or thermodynamics; a resolution of this issue may provide the field of semeiotics with some legitimacy and universality. In reviewing the field of information, Schalkwijk[10] states categorically that its "theory is a discipline different from physics. Whether or not there are important connections between these fields is not clear at present," and the "comparison has ... not yielded any results of practical importance." Similarly Wehrl,[11] in an extended review of properties of the function of entropy, suggests that "it is amusing to note that in practical applications of

---

[8] Tjalling C. Koopmans, *Three Essays on the State of Economic Science*, New York, 1957.

[9] J. L. Kelly Jr., "A New Interpretation of Information Rate," *Bell System Technical Journal*, **35**, 1956: 917-926, I. H. La Valle, "On Cash Equivalents and Information Evaluation Under Uncertainty," *Am. Stat. Assoc.*, **63**, 1968: 252-276, K. Arrow, "The value and Demand of Information," Chapter 12, in his Essays in the *Theory of Risk-Bearing*, Chicago, 1971: 267-278.

[10] J. P. M. Schalkwijk, "On a quantitative definition of information and its impact on the field of communications," *Rep. Prog. Phys.*, **45**, 1982:1213-1260.

[11] A. Wehrl, "General properties of entropy," *Rev. Mod. Phys.*, **50**, 2, 1978: 221-260.

information theory, such as technology, biology, etc., the second law of thermodynamics has been adopted and there called the negentropy principle. Thus," he continues, "we find a mutual interaction between physics and information theory rather than a perfect understanding of statistical mechanics on the grounds of information theory." But there are some important items that must be recognized: The principle that entropy is a measure of our ignorance about any physical system (and, reified, the semeiotic is such a system too) is well recognized since Boltzmann's time. Specifically, however, the mathematical theory of information of Weaver and Shannon [1943] deals with some very specific, yet abstract situations. In it, a message consists of words (in English, say) or digits. Both can be represented by a sequence of binary digitys (0s and 1s). Let this message require n digits for its representation. Then the set of all words with n digits each contains $N = 2^n$ elements so that $\log_2 N = n$. In general this is the measure of information. Consider now K sets E_s with N_i members for each E_i. Call $p_i = N_i / N$ where $N = \Sigma N_i$ For each member one then needs $\log_2 N_i$ pieces of information to identify it so that for the whole complex one would need

$$I = \sum (N_i/N) \log_2 N_i = \sum p_i \log_2 Np_i = \sum p_i \log_2 p_i + \log_2 N.$$

But note that $\log_2 N$ is the needed information when we have no data on the contents of each E_i The celebrated Shannon formula follows:

$$\log_2 N - I = \sum p_i \log_2 p_i$$

The connection to physics is obvious: if N is a set of measurements on a system, while p_i are the probabilities of finding the system in a pure state $|i\rangle$, then Shannon' s formula *coincides* with that of entropy.

The feeling of uneasiness with regard to the specifications of the information field is widespread. The mathematician J. L. Doob,[12] for instance, expressed his wonderment by asking whether it can be "that the existence of a mathematical basis [to information theory] is irrelevant, and the basic principle is the very idea that there is a context in which the word 'information' is accepted by general agreement and used in an intuitive way, and that no more is needed?" So one notices that specialists in the field describe it as mathematical to stress the fact that their work has no connection with reality and the physical world; in this manner, any expectations about utility and physics are immediately suppressed. Yet mathematicians reject it as not being one of their own. As Seidenfeld remarks: "thirty-six years after Shannon and Wiener introduced their familiar expression for the uncertainty captured in a probability distribution, entropy formalism is a thriving enterprise. Its advocates find applications in diverse settings..."[13] Formal information and communication theory in addition *appears* to be sterile. The reason is that both rest on some *ad hoc* assumptions about statistics and eschew any utilitarian content or any contact with the physical world. In fact, however, they surrepticiously address two fundamental economic problems: the one is the economic scarcity of semeiotic paths because

[12] In an editorial for the *IEEE Transactions of Information Theory*, March 1959.
[13] T. Seidenfeld, "Entropy and Philosophy," *Philos. of Sc.*, 53, 1986:467-491.

of noise, bandwidth and other physical constraints which have to be circumvented, and the other is the cost of management and the security of these channels which necessitates the overhead of "protocols" and "codes." In this connection I refer to the three theorems developed by Shannon which are the *guidelines* for information practice:

a. the SOURCE CODING theorem or data compression theorem which provides for the determination of the *shortest* string needed to specify an object,
b. the CHANNEL CODING theorem addressing the *reliability* of transmission of the semeiotic through a noisy, stochastic medium and,
c. the RATE DISTORTION theorem which addresses the kind of *distortion* that results from a truncated semeiotic.

The italics emphasize the value content of these theorems. As is the case of the second law of thermodynamics, these do not provide algorithms for reaching the desired results. But with their help, communication theory acquires a physical and economic content (by fiat rather than content). Indeed, noise is a basic constraint which information theory attempts to circumvent by developing tradeoffs: bandwidth vs power, and error-correcting codes. The constraint that is relevant here is the scarcity of channel capacities. Realism, and hence value, in the Shannon semeiotic is breathed in by considering it embedded in a physical world of noise and carried in time by some physical system (the channel) that relates to an ever evolving state of technology. Thus it is connected to the economic scarcity. As techniques advance, there is an apparently continuous innovative exploitation of physical possibilities to enlarge this capacity and thus obviate scarcity. An example of this is first the utilization of microwaves in free space and guided transmission paths and more recently of optical channels in dielectric guides (fibers). Some, in fact, ask the question: "what would you do with a zero-cost bandwidth?". Clearly, this is a non-economic question. There is no such thing as a zero-cost item. In fact, the increased bandwidth of the fibers, as such, is placing increased demands on its management.

Economics require the best possible utilization of the channel resources and here comes to the rescue the sampling theorem:[14] a continuous, analog semeiotic (voice, music, physical data from real-time event as in radar or an experiment) can be sampled at time intervals about equal to the inverse of twice the bandwidth it occupies and, then, be reconstructed with no observable differences from the original semeiotic! I note that of all the theoretical work in this field, the sampling theorem is thought to have had the greatest impact. This process of deconstruction and subsequent reconstruction of the semeiotic, motivated by the need for speech encipherment, permits, not only its eventual digitization, but specifically the intercalation and the packing of the semeiotics of different origins or pieces from the same origin. This approach along with the availability of a hierarchy of channels leads to better economic utilization (packing) of the channels. Value derives from the darwinian evolutionary mechanisms. Information is an economic concept, an evolutionary survival process that ex-

[14] Often attributed to C. Shannon, this theorem can be traced back at least to E. T. Whittaker, *Proc. Roy. Soc. Edin.*, **35**, 1915 if not to A. L. Cauchy in 1841; it was implicit in AT&T's own J. R. Carson in 1920 and H. Nyquist in 1928.

presses the inability of any construct to have a synoptic view of the universe. Information is a darwinian construct.

Semeiotics and Measurement. In the field of information theory, a "measurement" is also implied and *a priori* statistics assigned to a semeiotic. But, unlike the situation in the physical problem, there is no interaction with anything;[15] and there is no inherent dynamic whatsoever. Whatever happens to the semeiotics of information theory is exogenous. Shannon's approach is, thus, *ad hoc* and operational. The elements with which he operates do not make known anything other than themselves. There is no *renvoi* (feedback, in the sense of Jacobson) although codes can be developed which feedback the errors propagated in the transmission of a semeiotic.[16]

The macroscopic data, produced in the process of (any) measurement relating to thermodynamic extensive quantities, can be treated in the information theoretical sense. From such measurement processes that reveal noise and fluctuations, *a* microscopic world is conjectured to exist. This drives towards the reductionist efforts of statistical mechanics. It is only when, in reductionism, we seek the underlying reasons for these properties on the basis of "fundamentals," i.e. microscopics, that issues of subjectiv-

---

15 A. Katz, *Principles of Statistical Mechanics: The Information Theory Approach*, San Francisco, 1967. Katz provides a well argued presentation of the measurement in the two fields. However, to make contact with physics he introduces the Gibbsian ensemble from the back door.

16 One may consult a bibliographical essay oon semeiotics relating to information by T.A. Sebeok, *A Sign is Just a Sign*, Chapter 8 in paricular "The Evolution of Semiosis," Indiana University Press, 1991. But there we encounter phrases such as "Shannon's elegant, highly abstract, and therefore powerful theorems..."

ity vs. objectivity[17] crop up. The contact with "information theory" is not accidental. However, information theory makes no pretenses at being a physical theory. It is a statement in logic. It connects to the actual world only surreptitiously and never in a sharply articulated way. Statistical mechanics, on the other hand, purports to be a successfully reductionist theory relating to the physical world. Statistical mechanics has two aspects to it. The one relates to the equations of motion (classical or quantum). Unprocessed, these equations do not have 'an arrow of time' and therefore no contact with what is observed. The second relates to the shedding of information so as to end up with manageable equations; this process is a matter of expediency. To the stripped equations of motion, the Gibbs ansatz is applied and contact with thermodynamics is thus achieved. The *process* is information theoretical to the core. The gap, now, remains to connect this process not with entropy but to the fact that its measure changes, increasing globally.

There are scientists who want to endow "information" with a formality and physicality that is attractive but there is nothing supporting the view that there is a law of nature relating to information. In statistical mechanics one rejects *potential* "data" (some say information) so as to retain those pieces that are essential in developing a macroscopic picture in real time on which decisions can be made. The mind cannot cope with what it also believes to be very large and looks for uniformities and summaries. It is an economic operation. Nature does not seem to care as "it" processes all the data at all times and in real time: it represents

[17] K.C. Denbigh, and J.S. Denbigh, *Entropy in Relation to Incomplete Knowledge,* Cambridge University Press, 1985; also Leff and Rex (1990:109).

itself. The moment we want to escape from the "analog" description of nature (that is nature itself) then we resort to "cost-cutting" operations. Of all the potential data little survives in both theory and in experiment. Information, then, comes into the picture as a measure of what is, as a minimum, essential to develop a working theory and, with it, to predict the future events or retrodict events of the past.

We have been inundated with essays on the importance of information in all walks of life, including its very origins. The information society, cybernetics and control, DNA and information, order from chaos, and on and on. However, recognition of the obvious does not provide clues for the advancement of the field, does not provide support for policy decisions, and does not touch critically on the foundations of the semeiotic field. As discussed in Chapters II and III the basic economic inputs in a system are energy, materials, capital, and labor. All are scarce and there is a degree of substitution between them. Information is needed to manage all four inputs and is itself an overhead: it is transmitted and processed by utilizing these same scarce resources. In the Input-Output tables the sector is hidden in the Telecommunications Sector. Yet, one is informedthat there is an effort going on[18] to come up with something more substantial on the subject of information: to show that it is as basic as mass, energy, and the quantum. That these researchers are "crafting a new interpretation of quantum theory and quantum cosmology. Both are built on a scaffold of information." That "an electron is all information."

[18] By prominent researchers at Santa Fé; in the press and by G. Johnson, *Fire in the Mind: Science, Faith, and the Search for Order*, Alfred A. Knopf, 1955: 146, 159.

More ambiguous is the statement that "our task is to find an algorithm, a natural law that leads to the origin of information."[19] But more pragmatic is Dennett's point of view when he argues that "..the language of DNA and the 'readers' of that language have to evolve together; neither can work on its own. When deconstructionists say that the reader brings something to the text, they are saying something that applies just as surely to DNA as to poetry; and the something that the reader brings can be characterized most generally and abstractly as information, and only the combination of information from the code and the code-reading environment suffices to create an organism."[20] Indeed, Dennett expresses the need for many resources in a roundaboutedness to develop an organism. A position close to the one discussed in Chapter III.

The point we wish to make in the above commentary on "information" is that information acquires meaning in an environment of scarcity of resources needed to map the cosmos within the confines of a finite neurological-cognitive machine. Survival and biological operation necessitate the shedding of or the isolation from signals and combatting noise as needed. The signals themselves are physical entities which, if differentiable from environmental noise, are expressions of non-equilibrium states. Why then give information a privileged position in physics or biology when it, in fact, expresses the limitations of cognition? It is my contention that "information" is not a fundamental physical *existent*.

---

[19] M. Eigen, *Steps Towards Life,* Oxford University Press, 1992:12.
[20] D.C. Dennett, *Darwin's Dangerous Idea —Evolutin and the Meanings of Life*, Simon & Schuster, 1995: 196.

5. PROLEGOMENA ON MEASUREMENT

Introduction of Three Categories of Experiments

A narrative for an experiment cannot be very useful unless set in an appropriate framework that justifies the allocation of resources for its study. The analysis must be made in reference to its scientific merits and to its potential impact on issues that concern historians and philosophers of science. In attempting to position an experiment within the current thinking, however, one encounters ambiguities which prompted me to review the relevant literature and propose three categories in order to differentiate among "experiments." I summarize here on-going work. As noted in the introduction, this appendix is the setting of an agenda for research rather than a presentation of work reviewed and completed. The references are necessarily incomplete and names of authors are mentioned to provide a shorthand for association.

On scanning the literature on experiment within the above contex,t one is astonished at the lack of systematization indicating that the experiment-theory nexus needs more analysis than it has received this far. In this very preliminary reading of the literature, I have been baffled by a number of questions, some of which I list below.

- Is it taken for granted that all experiments are of the same nature? Is the cognitive aspect of object/pattern recognition in a microscope, or a radar scan of the Venusian surface of the same category as the study of parity conservation or decoherence?
- While all experiments in large laboratories may have the same characteristics in the formation ("construction", to use the current fad) of facts, do the objects of research have similar characteristics? Or is the early positioning of Kuhn[21] (Kuhn's "5th law" of thermodynamics?) a plausible avenue for describing the experimental cosmos?

[21] Thomas S. Kuhn, "The Function of Measurement in Modern Physical Science," *ISIS*, 52 (1961) : 161-190. (This article reflects the positions of text writers of the 50s and 60s whose concern has been Kelvin's dictum).

• Hacking[22] sets the tone of the inquiry by identifying the sequential steps in experimentation. The question is whether there are (within the sequence arrangements) experimental categories inherently separated by their physical-theoretical underpinning?
• Galison[23] provides a general view contrasting positivist and antipositivist schools with his "bricklayer's metaphor" in an attempt to identify "the mediative process by which experiments and theories contain each other." Similarly, Franklin[24] addresses this interactive process. Is it not that indirectly both address the mechanism of closure of [scientific] debates which apparently converge faster than other debates?[25]
• Is there a hidden agenda, a pecking order of importance?[26] Are some experiments more important than others? Although all experiments must be able to fit in a general theory one recognizes that, in fact, not all experiments have similar intellectual or practical weights.
• Experiment and passive observation are not often taken to be distinct. Is astronomical observation an experiment? How does it depend on instrumentation? In physics and engineering there is the concept of the "response function" which touches on the most sensitive cords of theory and experimental verification. It implies not an observation but rather an active wiggle of a system and an expected response,[27] i.e. "intervention."
• Why is there a cleavage between historians and philosophers of science? The latter are more atune to the physicist's concerns about the meaning of observation in a quantum mechanical sense. On this score historians are totally silent. The reason is not difficult to explain: the

[22] See Symposium, "The Philosophical Significance of Experimentation," *J. Phil.* LXXXV, Oct. 1988: 507-527. APA symposium, Dec. 28 1988: Ian Hacking, "On the Stability of the Laboratory Sciences," 507-514; Patrick A. Heelan, "Experiment & Theory. Constitution and Reality," 515-524. P. Galison, "Philosophy in the Laboratory," 525-527.
[23] See ftn. above.
[24] Allan Franklin, *The Neglect of Experiment,* California University Press, 1986.
[25] Harry N. Collins, "An Empirical Relativist Programme in the Sociology of Scientific Knowledge," in K.D. Knorr-Cetina & M. Mulkay, eds., *Science Observed: Perspectives on the Social Study of Science,* (London: Sage Publications, 1983 : 85-114). Michael Billig, *Arguing and Thinking: A Rhetorical Approach to Social Psychology,* (Cambridge: Cambridge University Press, 1987).
[26] Steven G. Brush, "Prediction and Theory Evaluation: Subatomic Particles," Institute of Advanced Study, Jan. 1993.
[27] Kubo (1966), Martin & Schwinger (1959).

EXPERIMENTAL CATEGORIES (Tentative Ideas)

I	II	III
SIMPLE, ONE-PARTICLE-LIKE; COMPLEX INSTRUMENTATION	COMPLEX INTERACTING PARTICLES; STABLE; BOOTSTRAPED COMPLEX INSTRUMENTATION*	SYNTHETIC SYSTEMS COMPLEX AND/OR BOOTSTRAPED SELF INSTRUMENTATION; STRUCTURES
Foundational Problems of Measurement Interaction space. "External Actors" Massive confining structures Instrumentation Inputs-Outputs	Interaction Space. "External Actors" Scattering and Scattering matrices Massive confining or self-delineating structures; strong couplings. Ditto (optical, electronic, computational) •	Operationally *as if though*: Cartesian Dualism: Observer vs Object or Self.* Virtual Experiments. Origins of . . . Macro structures (engines, cells) Counterfactuals
A "simplistic" model (ex.: the "one-particle" Dirac Hamiltonian; space-time; unification models; prediction [BRUSH ftn. 26] Hamiltonians of Everything (HOE) without most of what is included in Category II and III!	Hamiltonian of Everything (HOE) • Judgmental-Critical Selection of Hamiltonian Elements. Rejections. • Equation(s) of Motion. "Many-body" theories. • Simplifications, n-particle correlations • Response Functions	III a. Direct Macroscopic observationss III b. Indirect
<u>Theoretical outcomes</u>: kT-independent Data: kT-dependent through instrumentation. Dependence on Cat. II but mostly and often on III (the Volta experiment, say!) i.e. electronics, band gaps, cooper pairs, Statistical Mechanics and Information theory (i.e. economics)	<u>Theoretical outcomes</u>: kT-dependent & Fluctuations Data: kT-dependent (inherently and through instrumentation) Inputs from Statistical Mechanics and Thermodynamics Instrumentation: dependent on same category of ensembles (II) i.e commensurable with object to be measured	←(contributions to I and II) Information-theoretical & Statistical concepts; decision making

Examples: Millikan oil drops; Lamb-Retherford, Stern-Gerlach, Davidson-Germer experiments; Parity conservation; neutral currents; Fission and fusion; space-time. Also basic gravitation; large single macro-objects (mechanics, optics...) Phenomenological: Axiomatic Thermodynamics (?)** Hydrodynamics; cl. &rel. electrodynamics, mechanics	Examples: A. Chemical binding; elements of biological structures (DNA, RNA, etc) B. Solid state; gaseous and liquid structures; polycrystalline structures; astrophysical entities and spaces; C. Most instrumentation systems; Theories governing apparatuses D. Engineering. Praxis. Biotechnology. Dealing with complexity Scientific Impact: Present work Economic & Social Impact: histories	Examples: Planetary surfaces, cells, "Living Systems" (Replicative, mutating, unstable, evolution; origins of . . .), The brain structure, Cognition and the Mind; Thermodynamics (the economics of)
PHILOSOPHICAL STANCES: Quantum Mechanical theory of measurement. Decoupling. Heisenberg cut, projection operators, starting the von Neumann Chain* Theory vs Experiment as seen from III (positivism, antipositivism; hybrid) Mechanics of closure: rapid Objects of interest: refer to the many detailed accounts of various experimental episodes." [FRANKLIN, ftn 23].	PHILOSOPHICAL STANCES Duhem-Quine (applicable?); Aristotle's "rectification"? [*Nic.Eth.*] Scientific argument closure: not always rapid Mild Quintilian effects* Objects of interest: bubble chambers, planetary atmospheres, ionospheres and plasmas, electronic tubes, solid state devices (photo-devices, transistors, superconducting, etc) Covered by historians of Science	PHILOSOPHICAL STANCES Praxis, Dialectics, the tail of the "Von Neumann Chain"; mild forms of positivism vs antipositivism; metonymy; closure of debates: Quintilian-like; satisficing [Simon]; Theory making; intentionality of experiment; theory = a "virtual" experiment? Hacking's microscope studies: Object/pattern recognition (from cells to terrain, to facial expressions-body language; semiotics, drosophila); anthropology; economics & rhetoric; Hayek (1964): "complex phenomena", etc.
* On these and related issues see FTN. 29. ** Thermodynamics is a very idiosyncratic field.	* See ftn. 32.	* No position taken on this very controversial subject. "living instrument"!

TABLE A-1. Experiments are divided in three broad-gauged categories. For all categories instrumentation seems to develop from researches in II. Without II and (of course) III, I is meaningless. In III I include the cognitive condition in which also the philosophical questions are developed. I is of interest to III and interacts fundamentally with it. For it is with the synthetic systems operating on the the basis of II that I acquires content. In another essay I have worked the connection of II and III and have determined certain basic conditions which relate to thermodynamics. At the same time it allows for a model in which should III be an independent entity then the Cartesian dualism separates sharply the observed from the observer.

situation is too fluid and under considerable controversy to be submitted to a historical analysis. The only reference which comes close to this issue is the article by Heelan;[28] indirectly, the questions asked by Lockwood, Deutsch and, more recently, by Lelas[29] are relevant. For the physics community's approach to "measurement" see the Wheeler-Zurek collection of reprints[30]

These are some of the questions that seem not to have been addressed in the extant literature. To fill parts of the gap, I developed a tentative hierarchy of experimental situations with the tacit understanding that the categories cannot be sharply delineated. I perceive and list three distinct categories shown in the table on the next two pages.

Category I. It relates to the foundations of physics, i.e. the world of the quantum and of the particles. Calculations at this level do not, as they cannot, include statistical-thermodynamic concepts. This is recognized on perusal of basic texts as well as of advanced research papers. While temperature may affect a measurement, (non-quantum) statistics are irrelevant for the theory. Yet the apparatuses most likely constructed from principles understood in Category II do have important statistical inputs: signal in noise being the principal one. This affects the smoothness of the data. Theory gives a continuum; experiment discrete values.

Category II. The second category relates to the area which couples the fundamental to the cognitive. The theoretical underpinning of Category II is provided by I (reductionism). The subjects are systems which can be characterized with a Hamiltonian that describes all interactions between participating elements. Characteristic of this category is the assembly of strongly interacting elements for study and instrumentation utilization. It uses complex systems (massive "boxes," non-equilibrium devices, generators and amplifiers) to intermediate happenings at the microlevel with the cognitive, inquiring intellect. This category bridges the microscopic with the phenomenological, macroscopic, world.

---

[28] See ftn. 2, *Symposium*, 1988 : 515-524.

[29] Respectively: Michael, Lockwood, *Mind, Brain & the Quantum*, Oxford UK & Cambridge, Blackwell USA, 1991; David Deutsch, "Quantum Theory as a universal Physical Theory," *Int. J. Theor. Phys.* 24, 1, 1985 : 1-41; Lelas Srdjan, "Science as Technology," *Brit. J. Phil.* **44**, 1993 : 421-441. This is the first article that addresses some of the questions posed in this section.

[30] ***Quantum Theory of Measurement,*** Princeton University Press, 1983. Also *IBM JR&D* 32, 1988 and other work by Wheeler and Zurek.

Category IIb. It is possible to split Category II in two parts: IIa and IIb. In IIa one may retain what is already there. In IIb I consider observations and experiments that cannot fit the "response function" rubric: neutron stars, intergalactic plasmas, solar core, etc. Here models and hypotheses are invoked; we, as a rule, have data in need of Hamiltonians, as it were.

Category III. The third category is the synthetic. Here objects have purely macroscopic properties deriving from a multitude of microscopic systems and components. They consist of complex structures such as planetary surfaces, non-equilibrium situations, also structures that can replicate and mutate. The brain and the mind belong to this latter category. Cognition, object and pattern recognition and correlation [used to study categories I and II] are key characteristics.

Comments on the Table:

This table provides a tentative outline to be studied and, if necessary, rejected. The boxed item in the Category II requires further discussion. The way Hamiltonians are devised (or guessed) is an art which has yet to preoccupy the philosophers of science. One can formally write (better yet, guess) a "microscopic" Hamiltonian which will describe the energy of an assembly. From this, equations of motion can be derived and studied for consequences. But the equations of motion do not suffice to describe every macroscopic measurable effect of the system. For instance, given the Hamiltonian it is unlikely that conductivity — let alone superconductivity — would have been an *a priori* predicted behaviour of a system. In fact, experiment has been a consistent and a valuable guide from the beginning in the development of various notions on conductivity and superconductivity, leading finally to the breakthrough of Fermi statistics for conductivity and of Bardeen, Cooper & Schriffer electron pairs for superconductivity. Knowing the answer forces the investigator to specialize a Hamiltonian (pull a rabbit out of the hat, as it were) and its formalism from which the desired results can be obtained.

Whatever the quality of a theory in the Category I, it must fit the concept of the Heisenberg cut. This immediately allows for a decorrelation of the "external" from the "internal" with the concomitant rejection of components of a (presumed) complete theory. In Category II the situation becomes critical. Here, a Hamiltonian for the region under investigation — and the region may consist of a solid state chip or an ionospheric volume of many km^3 — can be written down ostensibly covering every possible interaction and particle property. However, only by specializing and weaning can one come close to a Hamiltonian and the derived equations of motion pertaining to the contrived or observed experimental situation. In this instance the Duhem-Quine[31] approach becomes relevant. Further, both theory and experiment can lead to novelty (inventive activity).

There is a pecking order of experiments. Some experiments deal with (what is thought to be) the foundations of physics: particle physics. But the vast majority of experiments deal with the world of established physics: the electron, the proton and the neutron, gravity, and the infinite variety of their combinations. Experiment is not only done for the study of natural phenomena but also in the "dialectic" mode of innovation.

When I refer to experiment, for lack of a more expedient route and in view of the above, I follow the path traced by the positivists, the antipositivists and those in between. A key element in all this is the assumption that theory — a derivative of the mind, itself of material content — can be decorrelated from the instrumentation and the object of measurement (the virtual Cartesian observer). This model is a working rather than a true model; any other route would have precluded any progress. Category I is meaningful in that instrumentation of Category II is thought to elicit responses from microscopic "objects" that are never accessible.

Discussion

• It is presently thought that we have a pretty accurate theory for many areas in physics. The theory, however, is useless without a variety of *ad hoc* and judgmental simplifications which we think make the theory approximate an experimental condition. For each experimental condition only a small part of the theoretical apparatus need be retained. Thus value judgments operate to save time and avoid unecessary complexity.

---

[31] For which see Peter Galison, "History, Philosophy, and the Central Metaphor," *Science in Context*, **2**, 1 1988: 197-212.

• The philosophical questions are no less demanding and obscure. Depending on the configuration of the many particle systems of which the systems are composed we may look into the subject matter from the quantum mechanical point of view and relax it to the classical picture where simplifications justify it.
• It is not a matter of theory or experiment first: The theory in II (classical or quantum) ostensibly exists. However, the theory of complex systems (the many particle systems) is a very intricate construct. Interactions with "external" systems — the process of measurement, in fact — are themselves complex and are functions of various space-time variables. The problem, then, is that it is not possible to visualize and predict every potentially realisable experimental virtuality. Some observations may even turn out accidentally. There is no algorithm that will list all possible experiments and/or outcomes. In this conext we may paraphrase "Quintilian's Principle of Uncertainty ('we can never capture the infinite variety of experimental situations [human affairs] in a finite system of algorithms [psychology, laws.]')."[32]
• We may wonder: can the Duhem-Quine thesis be studied in the context of Category II? While in the search for foundations we look for teleological unification, the synthetic world (Category III) has no such constraints: it is open ended, i.e. "dialectical."

Experiment to be connected with theory requires a heavy dose of statistical mechanics and thermodynamics. This is crucial because they are the steps taken to couple theoretical possibilities to capabilities of measurement within a finite time. This is rarely, if at all, discussed in physics texts or philosophical works. Thermodynamics and Statistical Mechanics appear to be extraneous constraints which, nevertheless, address the "economics" of measurement, i.e. the rejection of information rather than its accumulation.

---

[32] See Michael, Billig, *Arguing and Thinking: A Rhetorical Approach to Social Psychology,* (Cambridge: Cambridge University Press, 1987): compare with the Aristotelian "Equity" and Herbert Simon's "satisficing" [ftn. 43 chap. II].

BIBLIOGRAPHY

(The list contains some of the books and articles consulted; not every listing is referred to in the text and not every reference in the text is listed here)

ARON, R., *History of the Dialectic of Violence. An Analysis of Sartre's Critique de la Raison Dialéctique*, Trans. B. Cooper, Harper Torchbooks, 1975.

BARBIERY, M., "The Ribotype Theory of the Origin of Life," *J. Theor. Biol.,* **91,** 1981:545.

BERNSTEIN, H., *et al.*, "The Darwinian Dynamic," *Q. Rev. of Biology,* **58**, 1983: 185-207.

BERRY, R.S., SALAMON, P. AND C. HEAL, "On the Relation Between Economic and Thermodynamic Optima," *Resources and Energy*, **1**, 1978: 125.

BLAUG, M., *The Methodology of Economics*, Cambridge University Press 1980.

BLISS C.J., *Capital Theory and the Distribution of Income*, Amsterdam, North Holland, 1975.

BL[I]UMENFELD, L.A., "The Physical Aspects of Energy Transduction in Biological Systems," *Quarterly Review of Biophysics,* **3**, 1978: 251.

———, *Probleme der Molekularen Biophysik*, Berlin, 1977.

———, *Problems of Biological Physics,* Springer, 1981.

BÖHM VON BAWERK, E.R., *Capital and Interest,* Trans. G.D. Huncke and H.C. Sennholts, Libertarian Press, 1959.

———, *The Positive Theory of Capital*, trans. W. Smart, New York, Stechert, undated (1890?).

BRODA, E., *The Evolution of Bioenergetic Processes*, Pergamon, Oxford, 1975.

BUNGE, M., *Foundations of Physics*, Springer-Verlag, 1967.

BURMEISTER, E. AND R.A. DOBELL, *Mathematical Theory of Economic Growth,* Macmillan, 1970.

CALLEN, H.B., *Thermodynamics*, Wiley, 1960. Ch.4. "Processes and Thermodynamic Engines."

CAO, T.Y. AND S.S. SCHWEBER, "The Conceptual Foundations and the Philosophical Aspects of Renormalization Theory," *Synthese* 97, 1993.

CARATHÉODORY, C., "Grundlagen der Thermodynamic," *Math. Ann.* **67**, 1909: 355.

———, *Stz. Preuss. Akad. Wiss.,* 1925: 35-37.

CARNAP, R., *Two Essays on Entropy*, Ed. by A. Shimony, University of California Press 1978.

DAWKINS, R., *The Selfish Gene*, Oxford University Press, 1976.

DEPEW D.J. AND B.H. WEBER, *Darwinism Evolving —Systems Dynamics and the Genealogy of Natural Selection*, MIT Press 1995.

DICKERSON, R.E., "Chemical Evolution and the Origin of Life," *Sc. Am.*, **239**, 3, 1978: 76.

DOSE, K., "Prebiotic Polypeptides and the Origin of Biological Information," *Naturwissenschaften*, **70**, 1983: 378-382.

DYSON, F.J., "Time without End: Physics and Biology in an Open Universe," *Rev. Mod. Phys.*, **51**, 3, 1979: 447-460.

DYSON, F.J., Letter, *Sc.Am.* and response by Eigen, June 1981: 6.

EICHNER, A.S., ed. *A Guide to Post-Keynesian Economics*, M. E. Sharp Inc. NY 1978.

EIGEN, M., "Self-organization of Matter and the Evolution of Biological Macromolrcules," *Die Naturwissenschaften*, **58**, 1971: 465-532.

———, "The Origins of Genetic Information," *Sc. Am.*, **242**, 1981: 73.

———, "Selection and Evolution Governed by Natural Law," paper presented at the centennial of Darwin's death in Cambridge, 1982.

EIGEN, M. AND P. SCHUSTER, "The Hypercycle: A Principle of Natural Self-Organization. Part A. Emergence of the Hypercycle; Part B: The Abstract Hypercycle; Part C: The Realistic Hypercycle." Respectively in *Die Naturwissenschaften* of **64**, 11, 1977: 541-556; **65**, 7, 1978: 7-41; **65**, 7, 1978: 341-369.

EIGEN, M. AND R. WINKLER-OSWATITSCH, "The Transfer RNA, and Early Genes, *Die Naturwissenschaften* , **68**, 1981: 282-292.

EISENBERG, E. AND AND T.L. HILL, "Muscle Contraction and Free Energy Transduction in Biological Systems," *Science*, **227**, 1985: 999.

ELREDGE, N., "Information, Economics, and Evolution," *Ann. Rev. Ecol. & Syst.*, **17**, 1986: 351-369.

FISHER, I., *The Theory of Interest*, Macmillan, 1930 (Sentry Press, 1965)

FOX, S.W., *The Origin of Prebiological Systems and their Molecular Matrices*, NY Academic Press, 1965.

GEORGE, C., I. PRIGOGINE, AND L. ROSENFELD, *The Macroscopic Level of Quantum Mechanics*, Köngelige Danske Videnskabers Selskab, Mat-fys. Meddelesser, **38**, (12), 1972: 1-44.

GEORGESCU-ROEGEN, N., *The Entropy Law and the Economic Process*, Harvard University, 1971.

GERSCHENKRON A., *Economic Backwardness in Historical Perspective,* Harvard University Press, 1962

GIBBS, W.J,, *The Collected Works of J. Willard Gibbs*, NY, Longmans Green, 1931 (1920). *Elementary Principles of Statistical Mechanics,*, Yale University Press 1902.

GLANSDORF,.AND I. PRIGOGINE, *Thermodynamic Theory of Structure, Stability and Fluctuations*, NY 1971.

GRAY, B.F., "Reversibility and Biological Machines," *Nature*, **253**, 1975: 436.

GRAY, B.F. AND I. GONDA, "The Sliding Filament Model of Muscle Contraction, I. Quantum Mechanical Formalism" and, II. "The energetic and Dynamical Predictions of a Quantum Mechanical Trans ducer Model," *J. Theor. Biology,* **69**, 1977: 167 & 187 respectively.

GRÜNBAUM A., "Popper on Irreversibility," in *The Critical Approach to Science and Phiilosophy*, M.Bunge ed. in honor of Carl Popper, The Free Press of Glencoe, 1964.

GRASSMANN G., "Thermodynamik des Lebens aus der Blickwinkel der Technischen Thermodynamik und die Exergie," *Die Naturwissenschaften*, **71**, 1984: 335-341.

HAAG, R., *Local Quantum Physics: Fields, Particles, Algebras,* Springer-Verlag, New York 1992.———

HAYEK, F.A. VON, *The Counter Revolution of Science—Studies int he Abuse of Reason,* The Free Press of Glenco, Collier-Macmillan, 1955.

———, "The Theory of Complex Phenomena," Ch. 22 in *The Critical Approach to Science and Philosophy,* M.Bunge ed. in honor of Carl Popper, The Free Press of Glencoe, 1964.

HEISENBERG W., *Physics and Philosophy,* Harper and Row, NY 1958.

HINKLE, P.C. AND R.E. MCCARTHY, "How Cells Make ATP," *Sc. Am.* **238**, 1978: 104-123.

HIRSCHLEIFER, J., "Economics from a Biological Viewpoint," *The J. of Law and Econ.*, **XX**(1), 1977: 1-54.

———, "The Expanding Domain," *The Am. Rev.*, **76**, 6, 1985: 85.

HUDSON, E.A. AND D.W., JORGENSON, "U.S. Energy Policy and Economic Growth, 1975-2000," *Bell J. Econ. and Manag. Sc.,* June 1974.

INTRILLIGATOR, M.D., *Mathematical Optimization and Economic Theory*, Prentice-Hall, Englewood Cliffs, NJ, 1971.

ISRAEL, F. *Convexity in the Theory of Lattice Gases*, intr. by S. Wightman, Princeton, 1979.

JACOB, F, "Evolution and Tinkering," *Science*, **196**, 1977: 1161.

JACOB, F., *Logique du Vivant; une Histoire de l' Héredité*, Gallimard, Paris 1970.

JAMMER, M., *The Philosophy of Quantum Mechanics —The Interpretations of Quantum Mechanics in Historical Perspective*, Wiley 1974.

JOHANSEN, L., *Production Functions*, North Holland, 1972.

JUDSON, H.F., *The Eighth Day of Creation; Makers of the Revolution in Biology*, Simon and Schuster, NY 1979.

KEYNES, J.M., *General Theory of Employment, of Interest and, of Money*, London1936.

KILMEISTER, C.M., ed. *Schrödinger: Centenary Celebration of a Polymath*, Cambridge University Press, 1987.

KLEIN, M.J., "Maxwell, His Demon, and the Second Law of Thermodynamics," *Am. Sc.*, **58**, 1970: 84.

KOLM, S.-C., *Le Libéralisme Moderne*, Presses Univesitaires de France, 1983.

KOOPMANS, T.C., *Three Essays on the State of Economic Science*, McGraw-Hill, NY 1957.
———, "Intertemporal Distribution and 'Optimal' Aggregate Economic Growth," in W. Fellner, *Ten Economic Studies in the Tradition of Irving Fisher*, Wiley, NY 19..: 95-126.
KOOPMANS, T.C., "Ways of Looking at Future Economic Growth, resource and Energy Use," in M.S. Macrakis, ed. *Energy: Demand, Conservation and Institutional Problems*, MIT Press, 1974: 4-15.
KOVAČ, L., "Overview: Bioenergetics between Chemistry, Genetics, and Physics," *Rev. of Bioenergetics*, 1987: 331-372
KREGEL, J.A., *Theory of Capital*, Macmillan, 1976.
KRYLOV, N.S., *On the Foundation of Statistical Physics*, transl. A.B. Migdal, Ya.G. Sinai and Yu.L. Zeeman, Princeton University Press, Princeton, 1979.
KÜPPERS, B., "Towards an Experimental Analysis of Molecular self-Organization," *Die Naturwissenschaften*, **66**, 1979: 228-243.
LANDAU, L.D. AND M. LIFSHITZ, *Statistical Physiocs*, Pergamon 1958.
LEBOWITZ, J.L. AND O. PENROSE, "Modern Ergodic Theory," *Physics Today*, February 1973: 22.
LEFEBVRE, H., *Le Marxisme*, Presses Universitaires de France, Ed. *Que sais-je?*, 1961.
———,*Une Pensée Dévenue Monde*, Fayard, Paris 1980.
———, *La Présence et l' Absence*, Casterman, 1980.
LEFF, H.S. AND A.F. REX, "The Resource letter MD-l: Maxwell's Demon," *Am. J. Phys.* 58 (3), 1990: 201-209.
———, *Maxwell's Demon —Entropy, Information, Computing*, Princeton University Press, 1990.
LEIJONHUFVUD, A, *On Keynesian Economics and the Economics of Keynes*, Oxford University Press, 1968.
LEONTIEF, W.W., *The Structure of the American Economy 1919-1939*, Oxford University Press, 1951.
LEVI-STRAUSS, C., *The Savage Mind*, Weidenfeld & Nicolson, 1966.
LIEB, E.H., "The Stability of Matter," *Rev.Mod.Phys.*, **48**, 1976: 553.
LOTKA, A.J., *Elements, of Mathematical Biology*, Dover, 1956.
LWOFF, A., *Biological Order*, MIT Press, 1962.
LWOFF, A. AND ULMANN, eds. *Origins of molecular Biology; A Tribute to Jacques Monod*, Academic Press, NY 1979.
McCALL, J.J., "Probabilistic Microeconomics," *Bell J. of Econ. and Manag. Sc.* 1971: 403.
McCLARE, C.W.F., "Chemical Machines, Maxwell's Demon and Liv ing Organisms," *J. Theor. Biol.* **30**, 1971: 1 and **35**, 1972: 233.
———, "A Quantum Mechanical Muscle Model," *Nature*, **240**, 1972: 88.
MCCLOSKEY, D.N.,"The Rhetoric of Economics," *J. Econ. Lit.* XXI, 1983.
MACRAKIS, M.S., *Energy: Demand, Conservation, and Institutional Problems*, MIT Press, 1974.
MARGLIN, S., *Growth, Distribution, and Prices,* Cambridge, 1984.

MARX, K., *Capital*, English translation by S. Moore and E. Aveling, NY 1936; and Foreign Languages Publishing, Moscow 1961.
MASSIGNON, D., *Méchanique Statistique des Fluides*, Dunod, 1957.
MONOD, J., *Le Hasard et la Necécitée*, Editions du Seuil, Paris, 1970. Engl. edition: *Hance and Necessity —An Essay on the Natural Philosophy of Modern Biology*, Vintage, 1972.
PENROSE, O., *Foundations of Statistical Mechanics,* Pergamon Press 1970.
PENROSE, O., "Foundations of Statistical Mechanics," *Rep. Progr. Phys.,* **42**, 1979: 129.
PIAGET, J., "Les Courants de l'Epistémologie Scientifique Contemporaine," in *Logique et Connaissance Scientifique*, Piaget, J. ed. Enc. de la Pléiade, Gallimard 1967: 1224-1271.
———, J., *Le Structuralisme*, Presses Universitaires, Paris 1968.
PINDYCK, R.S., "Energy Price Increases and Macroeconomic policy," MIT Energy Lab. 79-061WP, 1980.
PINDYCK, R.S AND J.J., ROTENBERG, "Dynamic Factor Demands and the Effects of Energy Policy Shocks," *The Am. Ec. Rev.*, **73**, 5 1983: 1066-1079.
PIPPARD, A.B., *The Elements of Classical Thermodynamics*, Cambridge, University Press, 1957.
PLANCK, M., *Treatise on Themodynamics*, 3d Edition, transl. from the 7th edition (1921, 1922) Dover, NY 1945.
POPPER, K.R, *Logic of Scientific Discovery*, 1959, Harper & Row, 1965.
POPPER, K.R., *The Poverty of Historicism*, Harper, 1964.
PRIGOGINE, I. , *Non-Equilibrium Statistical Mechanics*, Interscience, 1962.
PRIGOGINE, I., "Dynamic Foundations of Thermodynamics and Statistical Mechanics," in *A Critical Review of Thermodynamics*, Ed. E.B., Stuart *et al.* Mono Book Corporation, Baltimore 1970.
———, "Time, Structure, and Fluctuations," *Science*, **201**, 1978: 777-785. (The Nobel Address).
PRIGOGINE, I. AND A.P. GRECOS, "Topics in Non-Equilibrium Statis tical Mechanics," in *Problemi dei Fondamenti della Fisica*, Bologna 1979, Corso LXXII, Varenna 1979.
PRIGOGINE, I. AND I. STENGER, *La Nouvelle Alliance. La Mátamor phose de la Science,* Gallimard, 1979.
RADNER, R., "Balanced Stochastic Growth at the Maximal Rate," *Zeitschrift für Nat. Ekon.*, Suppl. 1 1972: 39-52.
RAPPAPORT, D.J. and J.E. TURNER, "Economic Models in Ecology," *Science*, **195**, 1977: 367-373.
RAWLS, J., *A Theory of Justice*, Belknap, Harvard University Press, 1971.
READHEAD, M., *Incompleteness, Nonlocality, and Realism, — A Prolegomenon to the Philosophy of Quantum Mechanics,* Oxford University Press 1987.
RIZZO, M.J., *Disequilibrium and all that: An Introductory Essay, in Time, Uncertainty and Disequiiibrium*, Lexington Books, 1979.

ROBINSON, J., *Collected Economic Papers*, Vols. I-V, Cambridge, MA, The MIT Press, 1980.
ROSE, S., Ed. *Against Biological Determinism*, and *Towards a Libera tory Biology*, Allison & Busby, 1982.
SARTRE, J.P., *Being and Nothingness* , 1947: 469..
———, *Critique de la Raison Dialectique: (Précedé de Question de la Méthode)*, Tome I, *Théorie des Ensembles*, Paris, Gallimard, 1980. Of this, *Question de la Méthode* was translatedby H.E. Barnes as *S earch for Method*, Vintage 1971. An abridged version of the *Critique* has been translated by R.D. Laing and D.G. Cooper as *Reason and Violence*, Vintage, 1971.
SAMUELSON, P.A., *Foundations of Economic Analysis*, 2nd ed., Har vard University Press 1960.
SCHRÖDINGER, E., *What is Life? The Physical Aspects of the living Cell*, Cambridge,University Press 1960.
———, *Statistical Thermodynamics*, Cambridge University Press 1952.
———, *Mind and Matter*, Cambridge University Press 1959.
SCHUMPETER J.A., *Theory of Economic Development,* 1911 & Harvard University Press 1949.
SOLOW, R.M., "Technical Change and the Aggregate Production Function," *The Rev.Econ.&Stat.*, 1957: 312-320.
———, *Capital Theory and the Rate of Return*, North Holland, 1963.
STEINDL, J., *Random Processes and the Growth of Firms —A Study of the Pareto Law*, Hafner 1965.
STENT, G., *The Coming of the Golden Age*, The Natural History Press 1969.
VARELA, F.J. AND JEAN-DUPUY, *Understanding origins, Con temporary Views on the Origin of Life, Mind and Society.* Kluwer 1992.
TOLMAN, R.C., *The Principles of Statistical Mechanics*, Oxford University Press, 1938.
TOLTSTOY, I., *James Maxwell, A Biography*, The University of Chicago Press 1982.
TRUESDELL, C. AND S. BHARATHA, *The Concepts and Logic of Classical Thermodynamics as a Theory of Heat Engines. Rigorously Constructed upon the Foundation Laid by S. Carnot and F. Reech*, Springer Verlag, 1978.
VAN HOVE, L., "Quantum-Mechanical Perturbation giving rise to a statistical transport equations," *Physica*, **21**, 1955: 517.
VON NEUMANN, J., *Thermodynamik Quantenmechanischer Gesamtheiten*, 1927;
VON NEUMANN, J., *Mathematical Foundations of Quantum Mechanics,* Princeton University Press 1955. (Original German 1932).
WATSON, J.D., *Molecular Biology of the Gene*, 3d ed. W. Benjamin, 1977.
WEHRL, A., "General Properties of Entropy," *Rev. Mod. Phys.*, **50**, 1978: 2.
WEINBERG, S., *Dreams of Final Solution —The Search for he Fundamental Laws of Nature,* Pantheon 1992

WHEELER, J.A., "Frontiers of Physics," in Corso LXXII, *Problems in the Foundations of Physics*, G. Toraldo di Francia, ed. North Holland, 1979.
WHEELER, J.A. AND W.H. ZUREK, *Quantum Theory of Measurement*, Princeton University Press, 1983.
WIGNER, E., "On the Probability of the Existence of a Self-Replicating Unit," in Shils, ed. *The Logic of Personal Knowledge. Essays presented to Michael Polanyi in his seventieth birthday, 11 March 1961*, Glencoe, Free Press, 1961.
WILSON, S.S., "Sadi Carnot," *Sc. Am.*, **245**, Aug. 1981: 102.
WOESE, C.R., "Archaeobacteria," *Sc. Am.*, **245**, June 1981: 94.
ZUBAREV, D.N., *Non-Equilibrium Statistical Thermodynamics*, Consultants Bureau, 1974.
ZUBAREV, D.N. AND V.P. KALASHNIKOV, "The Derivation of Time-Irreversible Generalized Master Equation," *Physica*, **56**, 1971: 345.

In the process of writing this esssay I have consulted sources to the extent that time and ability to do so permitted in order to see with my own eyes the claims and their support. Secondary sources were also consulted avidly while I have not discounted the popular(ized) accounts that have a long and distinguished tradition. However, the competitive urges of the modern era for contracts and recognition have led to a style of writing that builds often unwarranted claims and expectations; these have to be checked carefully.

INDEX

ADP (Adenosine Diphosphate), 69
anthropic principle, 98
archaeology of fission, 157
Archilochus, 103
Aristophanes, *The Birds*, 94 and *passim*
Aristotle and *epieikeia* (equity), 53
Aristotle's *potentia*, 156
Aristotle, 12, 61n
Aron, R., 49, 55
Arrow, K., 86n, 197n
assembly brick-by-brick, 167
ATP (Adenosine Triphosphate), 69
ATP hydrolysis, 143
ATP-ADP cycle compared with the steam-water cycle, 140
Austrian theory of capital and quantum mechanics, 170
autonomy of biology, 16
availability (or exergy), 190, 191
axioms of economics, 45-46
backcattering of electrons, 164
bandwidth vs. power, 200
Beckett's *Not I*, 180n
Beckett, S. 11n 180, 180n
Berry, R..S., *et al.*, 126
Billig, M., 207
biological, capital, 10, 24; components, 142; engines as capital devices, 138; form, 89; order, 61; structures and capital theory, 11
biological work and microscopic engines, 138 and *passim*
biology, evolutionary, 1
bipolar thick filaments of myosin, 143
Bl[i]umenfeld, L.A., 139n, 145
Blaug, M., 27
Bliss, C.J., 19, 35, 37
Bohr, N., 82n; and X-ray measurements of DNA, 82; 143n
Boltzmann Equation, 107
Braudel, F., 78
bricolage atomique, 126
Brillouin, L., 122
Brinkmann, L., 103
Brown, M., 33n
Brune, M. *et al.* observations relating to the von Neumann conjecture, 160
Brush, S.G., 207
Bunge's *quantons* , 169
Burmeister, E. and R.A. Dobell, 19
Butler's aphorism, 73
calculus of variations, 5n
Callen, H., 2
Calvin, W. H., 116n
Cao, T.Y. and S.S. Schweber, 56, 107n, 153n, 168
capital (K), 9, 18; accumulation and economic growth, 48; accumulation and origin of; and origins of, 1, 8; and production theory, 11; as "strategic financial element," 36; and assembly in time, 130; at the cellular level, 81; concepts for thermodynamics, 93; fluctuations, 39;

reification, 38; reproduction and its stochastic nature, 29 theory, 1, 18; disembodied, 38; intensive in desalination plant, 194; and the focal point of economics, 174; and capital-like structures, 178
capital-like structures, formation of, 123
capsules, formation of, 81
Carathéodory, C., 98-99
Carnap, R., 3
Carnot cycle, 100
Carnot engine, 97, 162; the smallest, 145
Carnot S., 102
Carnot's contribution, 131
Cartesian, cellular machine, 88; dualism, 13, 14, 156; and practicoinert system, 88; and *res cogitans*, 15; and *res extensa*, 15; theater, 14n
cell as an engine, 138
cell-box assembly, 81, 84
cellular membrane, 63
Central Dogma and coefficients in the I-O tables, 66, 91
Central Dogma, 66, 66n, 74; simplistic, 90
Changeux, J.-P., 14n
Chatman, S., 161n
chemical potential, 188
chicken and egg paradox, 62, 85
Chomsky, N., 14, 14n
cilia and muscles, 68
Clausius, R., 94
coarse graining, 137
cognition and measurement, 206, 209
cognitive center, 154
Cohen, K., 130n
Cohen, R.S., 16
Colins, H.N., 207
collapse of the wavefunction, 155, 157
collective economic sea, 35
collective level in economics, 31
collective signals, 31
collective vs. independent behaviour, 32, 53
Commoner, B., 102
complementarity between capital and energy, 4, 4n, 25, 45n
Connes, A., 14n
consciousness, 11, 62, 82
constrained irreversibility, 96
consumption (C), 19
Copenhagen School, 152, 155, 179
creation and annihilation operators, 120
creation and annihilation physics, 3
Crick F.H.C. and J. Watson, 63
Critique de la Raison Dialectique (CRD), 41
Davisson C.J. and L.H. Germer, 163n
decoherence, 11, 159-160
Deeley, J.N., 196
Dehmelt, H., 165n
delta function, 109n
Democritus, 58,
demon (see also Maxwell), 95
Denbigh, K.C., 148, 202n
Dennett, D.C., 14n, 135n, 205
Dennett and multiple drafts, 153 153n,
density, macroscopic, $\varrho(t)$, 108
Depew D.J. and B.H. Weber, 89
Depew D.J., 16
depreciation rate, δ, 27
Descartes, 13, 14
Deutsch, D., 113, 170n, 210n
dialectical order, 48; reason, 55, 56
dialectics of biology, 72
Dickerson, R.E., 62
diergodicity, 12, 44
Dillon, G., 48n
Dirac, P.A.M,.112n, 130n
discounting, 28, 29
disease of matter, 11n
dissipation-fluctuation theorem, 3n
distribution function in phase space f(p,q,t), 108 and *passim*

distribution of income, 37
DNA (desoxyribonucleic acid), 66 and *passim*
DNA-RNA transactions, 71
Dobell, 19
Dobjansky, . ., 72
Doob, J.L., 199
Dorfman, R., 30, 30n, 36n
double helix DNA, 62
Duhem-Quine thesis, 212
durable equipment, 61, 175
Dyson, F., 119, 153n, 180
Eco, U., 174
economic equation of motion, 27
economic growth, 29
economic theory, 4
Edelman, G.M., and memory reentrant signalling, 153, 153n
efficiency in energy utilization, 79
egg generator, 135
Eigen, M., 63n, 71, 75, 75n, 78n, 100n, 205
Einstein's inspiration from thermodynamics, 98
Eisenberg E,. and T.L. Hill, 139n
elasticity of demand, 5, 23; of substitution and of complentarity, 45n; of substitution, 23
electromagnetism, 105
electron, nature of, 163
elementary 2×2 I-O matrix, 65
Elredge, N., 90
encapsulation, 61, 81
Energetische Weltanshaung (des Engenieurs), 103
Energy Crisis or Entropy Crisis, 103
energy (E),
energy conservation, 2
energy policy, 1, 4
energy theory of value, 78
energy, a protophysical concept, 127
energy-capital interactions, 178
engines, 1, 138 and *passim*
ensemble, 109
ensemble, grand canonical, 115
entropy change, 128
entropy, 128; expression for, 189
epieikeia (equity), 53
equation of motion in economics, 42, 46
equations of motion, 10
equity, 53
Evans, R.B., *et al.* on thermoeconomics, 194n
evolution of structures and reductionism, 119, 210
evolutionary biology, 1, 14
exergy, 193
existent, 161, 205
experiment vs. theory, 213
experiment, 13
experimental virtuality, 213
experiments, categories of, 206, 208-209, and construction aspects of, 162; and preparation of, 162
externalities, 91
Fagesrström, T., 16
Fermat's theorem, 5, 30, 114
Feynman sum over histories and definition of capital, 170n
filaments (of actin and myosin), 143
final demand, 68 and *passim*
Finer, J.T., 138n
first capital, 37
fission, 157
fluctuation-dissipation theorem, 3, 3n
fluctuations, 2, 39, 76
Fox, E., 62n
Fox, S.W., 81
Franklin, A., 207, 212
Galison, P., 155n, 207
Genette, G., 161n
genotype, 73
Georgescu-Roegen, N. and the irreducibility of the Second Law, 99
Georgescu-Roegen, N., 5
Gerlach, W. and O. Stern, 164n
Gerschenkron, A., 9, 9n, 37n, 43
Gibbs ansatz, 113-114, 185
Gibbs ensemble, 112, 147
Gibbs, 186

GNP (Gross National Product), 4
Gödel, K., 97n
Golinski, J., 48n
grand canonical ensemble, 141, 187
Grassmann, G., 140
Gray B.F. and I. Gonda, 145
Gray, B.F.,143
Gregory, R.L., 135n
guidelines for information practice: source coding, channel coding, rate distortion, 200
Haag, R., 114, 156n. 158n
Hacking, I., 155n, 207
Hacking, I., 207n
Hamiltonians, guessed, 211
Hamiltonian, 30; external, 167
happenings, 156
Harrod, R.F., 29
Hawking, S.W. and W. Israel, 171n
Hawking, S.W., 136n
Hayek, F.A., von, 24, 24n, 51, 101, 170n
Heelan, P.A., 155n, 210
Heisenberg cut, 156, 209
Heisenberg's objective tendencies, 155
Heisenberg's ontology, 155
Hesiod and the Iron Age, 10
Hesiod, 10
Hesiodian economy, 10
Hirschleifer, J., 16
Holton, G., 151n
Hudson E.A., and D.W. Jorgenson, 26, 65
Huxley's Z-plates, 142
Huxley, H.E., 143
Huygen's principle, 30
I-O table 90, 91,
idealism, 55
incremental evolution (see progenot model), 77
incremental work in physics, 151
information is a value-ladden concept, 86
information theory of value, 11, 78
initiation of events, 120
Input-Output (I-O) table, 20, 21, 64 and *passim*
input-output activity, 43
instrumentation box, 162
interaction of microscopic with macroscopic systems, 158
internal energy, 186
interutility, 12, 53
Intrilligator, M.D., 19
investment (I), 19
Iron Age and scarcity, 10
irreversibility and puzzle of, 2
Irreversibility, early recognition, 116
irreversibility, 115
Israel, F., 104
Jacob, F., 104
Jaffé, W., 40
Johansen, L., 19
Joule experiment, 125n
Judson, H., 67n
Julian, emperor, 116n
Kant's *Critique of Pure Reason*, 50
Kant's foundations, 55
Kant's organisms: means and ends, 85
Katz, A., 202
Keller, E, F., 62n
Kelly, J.L., 197n
Kennan, J. 192n
Keynes', "animal spirits," 36; aphorism, 61
Keynes, J.M., 30-31
keynesion, 12, 50, 52
Klein, M., 94
KLEM (capital, labor, energy, materials), 22, 31
Klimontovich, Y.L., 107
KMS condition, 113
knitting operation, 121
Koopmans' theorem, 119n
Koopmans, 9, 29, 47, 118n, 197n
Kováč, L., 140, 147
Kregel, J.A., 33n
Krylof, N.S.,104
Kubo, R., 3, 3n, 114, 207n
Kuhn, T.S., 206

Küppers, B., 63
labor (L), 19
labour theory of value, 78
Landau damping, 111n
Landau L.D. and M. Lifshitz, *Statistical Physics,* 148n, 192
Landau, L.D., 177
Landau-Vlassov equations for plasmas, 111n
Landes, D.S., 37n
Lebowitz and Penrose, 113
LEED (Low Energy Electron Diffraction), 162, 163
Lefebvre, H., 53, 80n
Leff, H.S. and A.F. Rex, 95, 146, 148n
Lehmann-Haupt, C., 102
Leijonhufvud, A., 33n
Leontief, W., 21n
Levi-Strauss, C., 55
Lieb, E.H., 118
life, 1, and origins of, 1
living labor, 35
localism, 107
Locke's Doctrine of Signs, 196
Locke's three categories: physica, practica, semeiotike, 196
Lockwood, M., 153n, 210
Lotka, A.J., 78
Lwoff, A. and A. Ullman, 87n
Lwoff, A., 64, 64n, 160
Lydon J.E., 77n
machine elements in ultimate miniaturization, 142, 150
machines and size limitations, 139
Macrakis, M.S., 25n, 58n
macroeconomic production function, 19
macroscopic machines and development of work, 139 many-body economic theory, 52
March, J.G., 58n
margin, at the, 151n
marginal, 7
marginalist theory, 38
Marglin, S., 5n
Markov chain, 121
Markovian Postulate, 12
Marshall, A., 15
Marshall, I.N., 153n
Martin, P., and J. Schwinger, 114, 207n
Marx, K., 9, 36, 36n, 48
materials (M)
mathematical economics, 48
Maxwell's, equations, 104; unification of electromagnetism, 151n; materialism, 181
Maxwell, J.C., 5n, 94-95 103
Maxwellian demon, 122, 145, 146, 171, 177
Maxwellian demons inapplicable, 147
Mayer, E., 16
McClare, C.W.F., 143, 145
McCloskey, D.N., 37n, 51
means and ends, 85
measurement, and semeiotics, 202; in physics, 151; epistemology of, 153, 155; and prolegomena on, 206; and the process of, 13
membrane, permeable, 75
Merleau-Ponty, M. 172
Merton, R.C., 71n
microscopic engines and
biological work, 138
mind/body split, 13
molecular motors, plethora of, 150
Monod J., 74
Monod's *Chance nad Necessity*, 87
motility, 84
multiple drafts, 153
muscle elements, 143
muscle, 138
muscular activity, macroscopic, 144
muscular beginnings, 84
myosin, 143
naked gene, 62
narrator and narration, 161
narrators, scientific, 161
natural selection as a near
inevitability, 90
natural structural capital, 63

neo-classsical, 37
neo-darwinian synthesis, 76
neo-Keynesian, 37, 38
neo-Marxian, 37, 38
neoclassical economics, 19
neo-classical theory of capital, 19,
Nitzsche F., 97
noisy self-reproduction, 75
Nordhaus, W., 28n
normative economics, 24
North A.C.T., 77n
Nyquist's theorem, 114n
observables in economics, 36
observational states, 122
observer, the, 160
Oklo nuclear reactor, 156, 157
optimal control theory, 29, 30n
origin; of capital accumulation, 9; of capital, 173; of reproduction, 63; of capital, 17; of life, 1
Ostwald, W., 96
Pais, A., 6n, 164n
particle correlation functions, 2,111
passage techniques, 14, 15
Pauli Master Equation for transitions, 107
Penrose, O., 3n, 8, 99n, 122
Penrose, R., 136, 136n, 151, 153n
perpetual motion machines, 128
perpetuum mobile, 99
phase space, 109
phenomenological being, 76
phenomenology of irreversibility, 134
phenotype, 73, 85 and passim
philosophical schools, 57
physical capital, 10, 22, 34
Piaget, J., 57
Pindyck, R.S., 4n
Pippard, A.B.,. 192n
Planck M, 96 and *passim*, 97n, 127
Planck M. (*Treatise on Thermodynamics*), 96
Planck's, idealist illusions, 180; views on thermodynamics, 55, 178
plasmon, 50
Platt, D.E., 164
Poincaré, H., 6n
Poisson Equation, 110
Polya, G., 122
positive economic theory, 23
post-Hesiodic economics, 45
practicoinert economy, 69
praxis, 48
precapitalistic world, 39
Prigogine, I. and Grecos, A.P., 116
Prigogine, I., 4n, 103, 117
Process of the firts kind, 157
process of the second kind, 157
production feasibility frontier or Consumption-Investment frontier, 80
production theory, 11
progenot model see incremental evolution), 77
projection postulate, 157, 159
protobiont (the naked gene), 62
pull-resile cycle, 143
QFT (Quantum Field Theory), 3
quantons, 169
quantum enigma, the, 12
quantum mechanics and consciousness, 82
quantum-mechanical box-model of cell, 145
quasi-particle, 31
quasiactor, 32n., 31
quasicapital, 12, 52
Quine, W.V., 8, 8n, 175n
Quintilian principle of uncertainty, 213
Raison Dialectique (Sartre), 40
Ramsey, F.P., 28
Rawls, J., 29n, 50
Redhead, M., 56
reduction of the Second Law, 94
reductionism, 55, 56, 59, 95, 104, 106
reductionist, beliefs, 159; reductionist program in thermodynamics, 137
reentrant signalling, 153
reification in thermodynamics, 93

reintegrated disintegration, 41-46
Reiss, T.J., 14, 14n
renormalization in quantum electrodynamics, 153
replication, molecular, 168
replicative structures, 74
reproducibility, 135
reproduction, 64
resource economics, 5
retrodiction, 26
reversibility, 134
reversible equations, 26
rhetoric in economics, 6, 7n, 48n, 51
Robbins, L., 47, 47n
roundaboutedness, 64, 138
Saint-Simon, 102
Saint-Simonian school of planners, 101
sampling theorem, 201
Samuelson P.A., 5n
Sartre and scarcity, 44
Sartre's "practico-inert," 42
Sartre, J.-P., 41, 44n, 48, 49
satisficing, 4, 53
scarcity, 10-11, 24, 44, 45, 47, 58, 178, 199; and semeiotics, 196; in work, 126; of channels, 200; of resources, 26, 46
Schalkwijk, J.P.M., 197n
Scharf, D.C., 159, 159
Schelling, T.C., 52n
Schlipp, A., 99n
Schrödinger's *Statistical Thermodynamics*, 112; and *What is Life?* 6n, 62, 62n, 120
Schrödinger, E., 6n, 62n, 120, 172
Schumpeter's *creative destruction*, 49
Schumpeter, J.A., 6, 6n, 15, 43, 48, 49, 107
Schweber, S.S., 56
Schwinger's non-operator sources, 167
second quantization, 107n, 111
seed corn, 34
Seidenfeld, T., 199n
Seitz, F., 118n
semeiotics and measurement, 196 and *passim*
shaft work, 186
Shannon's formula, 198
Shannon, C., 196n
Simon, H. and satisficing, 53, 58, 213
Sinai, *et al.* (see Krylof)
Sinai, Ya.G., 104
Smith, J.M., 16, 89n
Smithian invisible hand, 71
smithion, 12, 50, 52
Snow, C.P., 103
Solow and his coefficient $A_s(t)$, 48
Solow, R., 4, 35
spin, a relativistic effect, 165
sponaneous encapsulation, 62, 76; and possibility due to quantum effects, 83
Spudich, J.A., 139n, 150n
SQA (Status Quo Ante), 132
Srdjan, L., 210
stability of matter, 118n
Stapp, H.P., 153n
state vector reduction, 155
Steindl, J., 71n, 121n
Stent, G., 78
Stern-Gerlach experiment, 162
stopping power, 125
Stuart, C.I.J., 154n
substitution vs. complementarity, 45n
substitution, 4, 4n, 5
superoperator, 4n, 123
surface to volume ratio of desalination plants and of intestines, 195
Swedberg, R., 6
Szilard box, 146
Szilard thought experiment, 146
Szilard, L., 122
teleonomic apparatus, 74
ter Haar, D., 177n
theory of capital, 17, 30
thermal non-equilibrium, 1
thermodynamic Green's Functions, 108n
thermodynamic Utility, 130

thermodynamics, and fluctuations, 2; and its legitimmacy, 2; as an offshoot of mechanics, 105; a utilitarian discipline, 127; foundations of, 1
thermoeconomics, 194
thought experiments, 151
Tipler, F.J., 170n
Tolstoy, I., 180
Tomaselli, S., 14n
Truesdell, C., 137
two- particle and higher distribution functions, 111
unification in physics, 5n, 59, 151n
unitary collision matrix, 121
utility function, 76
utility of consumption U(C), 28, 43
utility of replication U(R), 73
value, and information theory of, 15, 78; and energy theory of, 78; and labour theory of, 78; and thermodynamics, 15; in work, 68, 137
value, 15
virtual, capital, 12, 39, 52, 212; Cartesian observer, 12; Cartesian theater, 15; cognitive center, 154
virtualities becoming events, 162
virtualities, realization of, 152
visualization, 151
von Neumann chain, 13
von Neumann's process of the first and of the second kind, 157
von Neumann, J., 157
wasps, 101
Watson, J., 61, 73
wavefunction, ψ, 30; collapse, 155, 176
wealth (W), 40-41,
Weber, B.H., 16
Wehrl, A., 8, 95, 99n, 197n
Weinberg, S., 87n
Wetzmann's doctrine and the *central dogma*, 66
Weyl's tensor, 171n
What is Life? 6n, 62, 62n, 119
Wheeler J.A. and W.H. Zurek, 152n
Wheeler, J.A., 180, 180n
whence capital?, 9, 33
Wightman, A., 104
Wigner, E.P., 121, 121n
work, 68; defined, 125; in macroscopic machines, 139; muscular; value of, 137
working fluid, 132, 140
Yeats, W.B., 33
Yoxen, E.J., 62n
Zurek's quantum analysis of the Szilard box, 148
Zurek, W.H., 148n, 159n

MIX
Papier aus verantwortungsvollen Quellen
Paper from responsible sources
FSC® C105338

If you have any concerns about our products,
you can contact us on
ProductSafety@springernature.com

In case Publisher is established outside the EU,
the EU authorized representative is:
Springer Nature Customer Service Center GmbH
Europaplatz 3, 69115 Heidelberg, Germany

Printed by Libri Plureos GmbH
in Hamburg, Germany